FIRE- AND SMOKE-RESISTANT INTERIOR MATERIALS FOR COMMERCIAL TRANSPORT AIRCRAFT

Committee on Fire- and Smoke-Resistant Materials
for Commercial Aircraft Interiors
National Materials Advisory Board
Commission on Engineering and Technical Systems
National Research Council

Publication NMAB-477-1
National Academy Press
Washington, D.C. 1995

NOTICE: The project that is the subject of this report was approved by the Governing Board of the National Research Council, whose members are drawn from the councils of the National Academy of Sciences, the National Academy of Engineering, and the Institute of Medicine. The members of the committee responsible for the report were chosen for their special competencies and with regard for appropriate balance.

This report has been reviewed by a group other than the authors according to procedures approved by a Report Review Committee consisting of members of the National Academy of Sciences, the National Academy of Engineering, and the Institute of Medicine.

The National Academy of Sciences is a private, nonprofit, self-perpetuating society of distinguished scholars engaged in scientific and engineering research, dedicated to the furtherance of science and technology and to their use for the general welfare. Upon the authority of the charter granted to it by the Congress in 1863, the Academy has a mandate that requires it to advise the federal government on scientific and technical matters. Dr. Bruce M. Alberts is president of the National Academy of Sciences.

The National Academy of Engineering was established in 1964, under the charter of the National Academy of Sciences, as a parallel organization of outstanding engineers. It is autonomous in its administration and in the selection of its members, sharing with the National Academy of Sciences the responsibility for advising the federal government. The National Academy of Engineering also sponsors engineering programs aimed at meeting national needs, encourages education and research, and recognizes the superior achievements of engineers. Dr. Harold Liebowitz is president of the National Academy of Engineering.

The Institute of Medicine was established in 1970 by the National Academy of Sciences to secure the services of eminent members of appropriate professions in the examination of policy matters pertaining to the health of the public. The Institute acts under the responsibility given to the National Academy of Sciences by its congressional charter to be an adviser to the federal government and, upon its own initiative, to identify issues of medical care, research, and education. Dr. Kenneth I. Shine is president of the Institute of Medicine.

The National Research Council was organized by the National Academy of Sciences in 1916 to associate the broad community of science and technology with the Academy's purposes of furthering knowledge and advising the federal government. Functioning in accordance with general policies determined by the Academy, the Council has become the principal operating agency of both the National Academy of Sciences and the National Academy of Engineering in providing services to the government, the public, and the scientific and engineering communities. The Council is administered jointly by both Academies and the Institute of Medicine. Dr. Bruce M. Alberts and Dr. Harold Liebowitz are chairman and vice chairman, respectively, of the National Research Council.

This study by the National Materials Advisory Board was conducted under Grant No. FAA-93-G-033 with the U.S. Department of Transportation.

Library of Congress Catalog Card Number 95-73091

International Standard Book Number 0-309-05389-7

Available in limited supply from:
National Materials Advisory Board
2101 Constitution Avenue, NW
HA-262
Washington, D.C. 20418
202-334-3505

Additional copies are available for sale from:
National Academy Press
2101 Constitution Avenue, NW
Box 285
Washington, D.C. 20055
800-624-6242 or 202-334-3313 (in the Washington metropolitan area)

B701

Printed in the United States of America.

COMMITTEE ON FIRE- AND SMOKE-RESISTANT MATERIALS FOR COMMERCIAL AIRCRAFT INTERIORS

ELI M. PEARCE (Chair), Polytechnic University, Brooklyn, New York
BRUCE T. DeBONA, AlliedSignal, Morristown, New Jersey
FREDERICK L. DRYER, Princeton University, Princeton, New Jersey
HOWARD W. EMMONS, Harvard University, Cambridge, Massachusetts
TAKASHI KASHIWAGI, National Institute of Standards and Technology, Gaithersburg, Maryland
BARBARA C. LEVIN, National Institute of Standards and Technology, Gaithersburg, Maryland
JAMES E. McGRATH, Virginia Polytechnic Institute and State University, Blacksburg
JAMES M. PETERSON, Boeing Commercial Airplane Group, Seattle, Washington
PATRICIA A. TATEM, Naval Research Laboratory, Washington, D.C.

Technical Adviser

SCOTT CAMPBELL, Douglas Aircraft Company, Long Beach, California

Liaison Representatives

THOR EKLUND, Federal Aviation Administration, Atlantic City International Airport, New Jersey
RICHARD LYON, Federal Aviation Administration, Atlantic City International Airport, New Jersey

National Materials Advisory Board Staff

THOMAS E. MUNNS, Senior Program Officer
AIDA C. NEEL, Senior Project Assistant
JOHN A. HUGHES, Research Associate

NATIONAL MATERIALS ADVISORY BOARD

Acknowledgments

The hospitality shown by employees of Boeing Commercial Airplane Group in hosting the committee's site visit and the time and effort taken in meetings and presentations are greatly appreciated. Individuals who prepared presentations or conducted factory and assembly line tours included James Likes, chief engineer for Interiors, Everett Division; Bruce Alexander, Interiors Engineering; Alan Anderson, chief payloads engineer; John Daniels, Production Engineering; Robert Elliot, Technology and Product Development; Sally Hasselbrack, Materials and Process Engineering; Kenneth Fairchild, factory manager; Mark Fee, Interiors Design; and Peter Guard, Technology and Product Development. The committee's time at Boeing provided a practical view of what goes into the design and production of an aircraft interior.

The committee acknowledges Craig Egenes and Ken Dowd of Walter, Dorwin, Teague and Associates for their presentations on aircraft cabin architecture and interior design.

The committee also thanks the individuals who participated in the Conference on Fire- and Smoke-Resistant Materials. The information and ideas generated at the conference were invaluable to the committee's work. In particular, the effort that the invited speakers expended in preparing papers and presentations for the conference is appreciated. Papers were presented by Hans-Dieter Berg, Deutche Aerospace Airbus; Hanns-Jöerg Betz, Lufthansa German Airlines; Dougal D. Drysdale, University of Edinburgh; Sally A. Hasselbrack, Boeing Commercial Airplane Group; Thomas E. McSweeny, Federal Aviation Administration; Thomas M. Murray, Boeing Commercial Airplane Group; David Purser, Fire Research Station, U.K.; James G. Quintiere, University of Maryland; Constantine P. Sarkos, Federal Aviation Administration; Swen J. Schaich, Deutche Aerospace Airbus; Usman Sorathia, Naval Surface Warfare Center; Archie Tewarson, Factory Mutual Research Corporation; Edward D. Weil, Polytechnic University; Charles A. Wilkie, Marquette University; and Martel Zeldin, City University of New York. Also the committee acknowledges the efforts of the workshop session chairs: Stephanie Skaggs, New Mexico Engineering Research Institute; Richard Gann, National Institute of Standards and Technology; Dennis Nollen, DuPont; and John Rock, Amoco Performance Products.

The committee is particularly grateful to the Federal Aviation Administration liaison representatives, Thor Eklund and Richard Lyon, for their participation in a number of committee discussions.

The chair of the committee thanks the members for their dedication and patience during the preparation of the conference proceedings and the study report. Their diligence and goodwill made for an unusually effective team effort.

Finally, the committee gratefully acknowledges the support of Thomas E. Munns, National Materials Advisory Board senior program officer and Aida C. Neel, National Materials Advisory Board senior project assistant.

Preface

The Federal Aviation Administration (FAA) has established an Advanced Fire Safe Materials Research Program to meet the requirements of the Aviation Safety Research Act of 1988. The program's objective is "to discover the fundamental relationships between the composition and structure of materials and their behavior in fires to enable the design of a totally fire-resistant cabin for future commercial aircraft. Research will be basic in nature and will focus on synthesis, characterization, modeling, and processing of new materials and material combinations to improve the fire performance, increase the functionality, and reduce the cost of next-generation cabin materials."

The FAA requested that the National Research Council, through its National Materials Advisory Board, recommend long-term research directions in promising areas based on projected technology. Towards this end, the National Research Council established the Committee on Fire- and Smoke-Resistant Materials for Commercial Aircraft Interiors. This committee met several times during the course of its study and development of this report. The committee itself was balanced in regard to the varied science and technology aspects required for such an interdisciplinary study.

The committee hosted a Conference on Fire-and Smoke-Resistant Materials, which was held at the National Academy of Sciences on November 8–10, 1994. Conference attendees included representatives from industry, government, and academia. The conference participants identified trends in aircraft fire safety and suggested promising research directions for the FAA's program in smoke- and fire-resistant materials. The conference proceedings, containing 15 invited papers and summaries of the workshop sessions, have been published and serve as an important resource for the preparation of the committee's report.

The ultimate product of this committee report are the conclusions and recommendations on research opportunities that meet the requirements of the Aviation Safety Research Act of 1988. The committee strongly supports these conclusions and recommendations, and as experts in this area are very much aware of the high-priority urgency for support of these recommendations, even in these difficult times of "downsizing" research. A commitment to a longer-term view will allow us to have the necessary materials for fire safety for the new construction required for continued economical air transportation, as well as keeping the United States in its traditional leadership role in passenger aircraft construction.

Comments and suggestions that readers of this report wish to make can be sent via Internet electronic mail to nmab@nas.edu or by FAX to the National Materials Advisory Board (202) 334-3718.

Eli M. Pearce

Chair, Committee on Fire- and Smoke-Resistant Materials for Commercial Aircraft Interiors

Contents

	EXECUTIVE SUMMARY	1
1	INTRODUCTION	6
	Statement of the Problem,	6
	Accident Statistics,	6
	Statement of Objectives,	9
2	DESIGN AND FUNCTION REQUIREMENTS FOR AIRCRAFT INTERIOR MATERIALS	12
	Materials of Construction,	12
	Weight,	17
	Cost,	17
	Manufacturability and Producibility,	18
	Comfort,	19
	Maintainability,	20
	Aesthetics,	21
3	EVALUATION OF MATERIALS FIRE PERFORMANCE	22
	Fire Safety,	22
	Fire Scenarios,	23
	Fire Models,	26
	Testing and Fire-Safety Assessment,	27
4	DEVELOPMENT OF CANDIDATE MATERIALS FOR FUTURE INTERIORS	29
	Combustion of Polymers,	29
	Development of Materials,	30
	Manufacturing,	34
	Theoretical Modeling of Polymer Combustion,	36
5	CONCLUSIONS AND RECOMMENDATIONS: RESEARCH OPPORTUNITIES	39
	Materials,	40
	Component Design and Manufacturing,	40
	Fire Scenarios,	41
	REFERENCES	45
	APPENDICES	
A	GLOSSARY OF TERMS	51
B	FAA RESEARCH AND DEVELOPMENT	53
C	CURRENT FIRE-MODELING CAPABILITIES	61

D TOXICITY MODELING AND TESTING 65

E BIOGRAPHICAL SKETCHES OF COMMITTEE MEMBERS 69

List of Figures and Tables

FIGURES

1-1	Fatal accident rates and fatalities	7
1-2	Jet aircraft in service and annual departures,	8
1-3	Annual accident rates,	8
1-4	Causes of accidents in final approach and landing,	9
1-5	Fatalities classified by type of accident,	10
1-6	Primary causes of fatal accidents,	10
2-1	Aircraft interior cabin liner components,	12
2-2	Considerations in inclusive design criteria,	13
2-3	Typical cabin panel constructions,	14
2-4	Typical seat construction,	19
3-1	Summary of in-flight and on-ground smoke and fire accidents,	24

TABLES

1-1	Summary of Fatal Accidents,	9
2-1	Materials Applications in Current Commercial Aircraft,	15
2-2	FAA Flammability Requirements for Cabin Liners,	16
C-1	Characteristics and Limitations of Fire-Hazard Assessment Models,	63

Executive Summary

The Aviation Safety Research Act of 1988 (Public Law 100–591) provides the Federal Aviation Administration (FAA) with a mandate to conduct long-term investigations concerned with fire safety, including fire containment and the fire resistance of cabin materials. As part of their response to this legislation, the FAA Technical Center established a program to investigate improved fire-resistant materials for aircraft interiors, with the objective "to discover the fundamental relationships between the composition and structure of materials and their behavior in fires to enable the design of a totally fire-resistant cabin for future commercial aircraft."

The Committee on Fire-and Smoke-Resistant Materials for Commercial Aircraft Interiors was established to provide guidance for this effort. The two principal objectives of the committee's task were (1) to identify promising materials technologies, design issues (both overall and for individual components), and fire performance parameters (both full scale and for individual components) that, if properly optimized, would lead to improved fire and smoke resistance of materials and components used in aircraft interiors; and (2) to identify long-range research directions that hold the most promise for producing predictive modeling capability, new advanced materials, and the required product development to achieve totally fire-resistant interiors in future aircraft. The emphasis of the resulting study is on long-term innovation leading to impacts on fireworthiness of aircraft interiors 10–20 years hence.

CONCLUSIONS AND RECOMMENDATIONS

Aircraft interiors are complex systems that include visible items such as flooring, seats, lavatories, ceilings, sidewalls, stowage bins, bag racks, closets, and windows, and items that are not visible to passengers such as ducting, wiring, insulation blankets, and supporting structures. Polymeric materials are predominant, appearing in a wide range of product forms including molded sheet or shapes, composite-faced honeycomb sandwich, textile fibers (fabrics or carpets), foams, sealants, and adhesives. Interiors currently contain materials of varying fire resistance, selected for their particular application and a variety of additional factors such as availability, cost, producibility, and a balance of other useful properties.

The committee believes that long-term, focused research in fire-resistant polymeric materials can lead to significant improvements in fire performance and safety. To support the development of such materials, advances are required in the understanding and analytical modeling of aircraft fire scenarios, polymer combustion, small-scale characterization tests, and fire-hazard assessments. In addition to the materials' properties, the development process must address the needs of users of the materials. These needs include the processing and production capabilities of the materials suppliers and aircraft manufacturers; the ability to meet the design, performance, comfort, and aesthetic demands of aircraft interior applications; and compliance with environmental, health, and safety regulations and practices in the manufacture, use, and disposal of aircraft interior components.

The committee focus was on materials technology and enabling design, manufacturing, testing, and modeling capabilities. A comprehensive research program with the goal of improving survival of aircraft accidents would also include aspects of fire-safety and-suppression systems, human factors and behavior in emergencies, and sensor and control development for accident avoidance.

While the committee is confident that significant improvements can be realized in materials performance, the FAA goal of an "order-of-magnitude" improvement in fire resistance is difficult to define because of the multitude of performance metrics and fire scenarios that need to be considered and evaluated. Establishing specific performance goals for materials research based on the current understanding of materials combustion and aircraft fire scenarios is problematic because the data needed to relate materials performance and configurations to observed fire scenarios are not available.

- The committee recommends that materials performance goals for long-term research be established using hazard and risk assessment techniques. These techniques require experimental data from appropriate small-scale tests in conjunction with fire models to predict the expected fire performance and assess the probability of occurrence under realistic conditions, followed by validation tests in the intermediate-and full-scale regime.

Materials

Materials used in the production of current aircraft interiors, with some exceptions, tend to have better tire resistance than materials used in other transportation systems. Regulatory requirements have been significant driving forces in the optimization of fire-resistant polymers and the development of required product forms for aircraft interior applications. Independent programs pursued by industry have also resulted in essentially a new generation of materials that found application in the 747, DC-10, and L-1011, and then a second generation of materials used for the 767, 757, and A300-600/A310. The FAA's heat and smoke release regulations drove improvements to the second-generation materials and to the application of new materials such as more fire-resistant thermoplastics to satisfy specific application needs.

The committee has identified three approaches to provide further improvements in tire-resistant materials:

- Modification of specialty polymers including thermoplastics such as polyetheretherketone, polyetherimide, polyphenylene sulfide, and polysulfone and thermosets such as cyanate ester, bismaleimide, polyimide, and polybenzimidizole. This approach may provide the best performance in the near term ($\leq$ 10 years).
- Development of new, high-performance, thermally stable materials including organic/inorganic polymer systems, copolymers, polymer blends and alloys, and glasses and ceramics. These materials have the potential for the best performance in the long term (> 10 years).
- Modification of existing engineering polymers including thermoplastics such as polycarbonates, nylons, and polyethyleneterephthalate and thermosets such as phenolics and polyesters. While it is not clear that this approach would lead to the significant improvement in performance sought, this approach may result in significant cost reductions.

A basic scientific understanding of char and intumescence (swelling, foaming) is crucial to the development of improved materials. Research in char formation should include structural characterization and mechanical behavior (durability) and its relationships to ambient environment, heating rate, chemical derivatization, additives, and coatings.

The two general technical directions identified by the committee for polymeric materials development to improve fire and smoke resistance are incorporation of additives in polymers and synthesis of thermally stable, fire-resistant polymers. Particularly promising approaches are discussed in detail in Chapter 4. These include thin laminated or co-extruded films and blends, coatings and additives (including intumescents), phase-change or temperature sensitive materials, organic/inorganic polymer blends, polymer blends utilizing a high char-forming polymer as an additive, and polymer modifications. Additive approaches include volatile-phase active flame retardants that inhibit the combustion process, condensed-phase active flame retardant that lead to char or intumescence, flame retardants that endothermically lose volatile components, heat-sink additives, toxicant suppressants, and combinations of additives that take advantage of synergistic effects (i.e., multiple additives with differing but cooperative modes of activity in optimized combinations).

Recommendations:

- Perform research to improve the fundamental understanding of polymer combustion, including thermal degradation, char formation, intumescence, toxic gas production, and heat effects. Place special emphasis on the characterization of char and intumescence processes.
- Investigate new additive approaches that allow for significant improvements in fire resistance and reduced toxic gas production in current materials.
- Facilitate the development of new or modified polymers with significantly improved resistance to ignition and flame spread. Emphasize the modification of existing specialty polymers to obtain desired properties and the development of new thermally stable polymers or blends.
- Evaluate and prioritize research and technological development efforts to ensure that the new materials will meet end-use requirements. Issues to be considered include costs; the contemporaneous processing and production capabilities of the materials and aircraft industries; ability to meet the design, performance, comfort, and aesthetic demands of aircraft interior applications; and compliance with environmental and health and safety regulations and practices.

Component Design and Manufacturing

New fire-resistant materials are of little practical value for aircraft interior use if the industrial processing technologies required to manufacture parts are not fully developed and broad-based. Short-and long-term strategies should be developed to characterize new material opportunities for compatibility with existing processes, as well as determining needs for future designs and manufacturing technologies. Short-to mid-range strategies should focus on researching materials that can be produced with existing tooling and manufacturing processes. Long-term strategies should evaluate both materials that can be processed with today's technologies as well as with future technologies. Where improved tire performance can only be achieved with materials requiring new manufacturing processes, materials research and manufacturing process development should be conducted concurrently to ensure smooth implementation.

Innovative design and processing concepts could facilitate additional fire-resistant materials research options while allowing for reduced manufacturing costs. New modular design technologies should be pursued to reduce part count by integrating components. Minor changes to component designs may also yield improved fire performance. Lightweight films and coatings with improved fire resistance that can be easily integrated into current component constructions, such as interior sandwich panels or insulation blankets, should be investigated.

The development of manufacturing processes for future aircraft interiors will emphasize cost reduction. Process developments likely in future interior applications include increased automation of cutting and ply location processes; reduction in the number of process steps and decreased cycle times in molding and decorating processes; low-temperature-curing and quick-cure thermoset polymers; and net-(or near-net-) shape molding processes to reduce machining and trimming operations. Developing continuous-fiber-reinforced thermoplastic composites to take advantage of thermoplastic processing advantages could provide significant savings in manufacturing costs in future aircraft. Thermoplastic composites could offer the advantages of less expensive tooling, more versatile production methods, short production cycle times, elimination of hand finishing, increased durability, the ability to integrate decoration or texture, and the potential for recycling. Development of manufacturing processes and compatible materials for co-bonding, secondary bonding, and decorative processes are required before the full potential of thermoplastic composites can be realized.

Recommendations:

- Prioritize materials research opportunities in terms of compatibility with existing tooling and manufacturing processes. Short-to mid-range programs should focus on materials systems highly compatible with existing manufacturing technologies for a smoother introduction into production. In long-range development where new manufacturing processes are needed, materials research and manufacturing process development should be conducted concurrently to ensure smooth implementation.
- Investigate innovative design and processing concepts such as modular design, fire-resistant films and coatings, new thermoset composite materials and manufacturing approaches to reduce the number of processing steps and cycle times, and expanded use of thermoplastics. These concepts could provide improved fire resistance while reducing manufacturing costs.

Fire Testing, Evaluation, and Modeling

To perform an adequate fire-hazard analysis or flammability assessment of a material, several materials characteristics must be integrated. Experimental data from appropriate small-scale tests should be used in conjunction with fire models to predict the expected fire performance and the probability of occurrence under realistic conditions, followed by validation tests in intermediate-and full-scale regimes. This type of assessment is sensitive to the end-use application, type of material, and actual fire threats.

A complete understanding of aircraft fires and the responses of materials and components in these fires is required to establish appropriate performance goals and evaluation criteria for new fire-resistant materials. Based on prior experience, two basic aircraft fire scenarios have been identified: post-crash fires involving (potentially large) quantities of aviation fuel from ruptured fuel tanks and in-flight fires involving only interior cabin components and passenger-specific items. These scenarios provide the basis for establishing fire performance behavior and criteria for new materials. However, new aircraft configurations may be significantly different from past designs, and the response of aircraft interiors in these fire scenarios depends on the details of the design. Thus, each aircraft configuration must be analyzed to assess its response.

In the past, in-flight interior fires have very rarely developed into accident scenarios. Those fires within the passenger compartment have been detected and extinguished before posing a significant threat, and most that began in or around lavatories either were detected and extinguished or self-extinguished. Therefore, in-flight fires in accessible areas within the aircraft interior were not considered in this study. However, in flight fires in inaccessible areas can be a serious concern because of the potentially long periods (up to three hours) before passengers can be evacuated. The number of accidents that began as a fire in a cargo compartment is, relatively speaking, extremely small. Nevertheless, recent regulatory upgrades of cargo compartment liners require them to perform as a substantial fire barrier to contain possible fires from spreading into the passenger compartment.

Adaptation of current test methods, and, in many cases, new small-scale test methods, are needed to evaluate fire performance characteristics of materials for specified aircraft interior situations and to provide property data for use by modelers to predict component fire performance in expected large-scale fire scenarios. Rather than being used exclusively as pass/fail screening tests, small-scale tests should be used to measure flammability properties of the materials that can be used as input to theoretical models to predict fire hazard. This process requires enhanced interaction between the experimentalist and the modeler to establish that the test procedures are designed to obtain the parameters that the models require. A better understanding of the performance, limitations, and operating principles of existing test equipment and the development of new and better test methods are needed.

The development of a materials fire-test database would provide a framework to establish performance criteria, evaluate new materials, and predict materials behavior in-service. The steps involved in database development are (1) categorizing and cross-referencing test methods and characteristics, identifying the fire parameters needed for hazard assessment and modeling; (2) compiling existing fire characterization data; (3) obtaining relevant fire-test data through testing programs as they become available; and (4) developing new test methods where needed to provide additional data not available from current methods.

An integrated modeling capability for aircraft interior designers could allow the estimation of the performance benefit of various choices of fire-resistant materials and components in aircraft interior applications. Analytical models, ranging in scale from molecular to full scale, are needed to support the development and evaluation of new fire-resistant materials. Models could be used to predict materials performance in fires, to assess the fire hazard, and to establish performance goals for new materials.

Thermal degradation models that include crosslinking, cyclization, aromatization, and network formation could be used as tools to determine ways to enhance the thermal stability of polymers and to promote char formation during polymer degradation. Intumescent char models based on the formation and growth of bubbles, swelling, polymer melt behavior, and carbonization may be used as tools in optimization of intumescent materials or coatings. The models should be applicable to engineering polymers, specialty polymers, polymers with fire-retardant additives, and polymer blends.

To provide an understanding of how materials and components contribute to development of a full-scale fire, largerscale fire models need to be improved for the specific needs of aircraft interior designers. Fire-growth models, compartment fire models (zone and field models), hazard assessment models, and toxicity models need to be developed or modified to address the aircraft interior configuration and relevant fire scenarios.

Finally, the models under development must be validated to gain the confidence of the design community. This requires a closely coordinated research effort between theoretical model development and intermediate and large-scale validation testing. More emphasis needs to be placed on requiring intermediate and large-scale testing to verify small-scale data and to refine the effects of size and configuration on the fire performance.

Recommendations:

- Develop the science base for small-scale fire performance and toxicity tests, based on expected fire scenarios and verified with full-scale tests, to provide meaningful property data for modeling and materials evaluation.
- Develop a database of materials fire performance properties to provide a means to establish performance criteria, evaluate new materials, and predict materials behavior in aircraft applications correlatable with expected fire scenarios.
- Support technology scale-up through testing on an increasing scale, from small-scale through full-scale testing.
- Develop basic thermal degradation models that are applicable to engineering and specialty polymers and include crosslinking, cyclization, aromatization, and network formation to aid the understanding of polymer stability and char formation. Include both char characteristics and evolved gaseous product properties as key model parameters.
- Develop intumescent char models based on the formation and growth of bubbles, swelling, polymer melt behavior, and carbonization.
- Develop an integrated modeling capability that will allow the estimation of the performance benefit of various choices of fire-resistant materials and components in aircraft interior applications. Work is needed to develop fire-growth, toxicity, and hazard assessment models relating to aircraft fire scenarios.

Long-Term Research Program

The committee believes that the goals of the FAA's research program to develop order-of-magnitude improvements in materials fire performance cannot be met with incremental advances or near-term regulatory activity. Rather, what is required are substantial advances based on a fundamental understanding of polymer combustion, on accurate aircraft fire scenarios, and on the *systematic* development of materials technology improvements. These advances in turn require a long-term commitment on the part of the FAA working with the aircraft and materials industries and research laboratories.

The uncertainty of new commercial programs, the cost of qualification and certification, and the long time-to-market for new materials tends to discourage suppliers from embarking on materials development efforts. The size of the potential market for materials for use in aircraft interior components often does not justify the expense to the suppliers of development and qualification. Thus, it is important to develop alternate markets for new materials and to apply technological developments from other industries.

Many of the developments that arise from this research will be unique to the issue of commercial aircraft interior fire safety. However, advances in the understanding of polymer combustion, new materials and additive technology, and testing and modeling methods may have applicability to fire safety in other transportation systems such as ships, trains, automobiles, and buses, as well as commercial and residential buildings. If this long-term research effort is sustained and a coordinated, parallel effort persists in these related areas, significant advances will be made in the understanding of materials fire safety not only for commercial aircraft interiors but also in many other areas where fire safety is a concern.

Recommendations:

- Sustain the effort to develop significantly improved fire-resistant materials as a long-term research program, with clearly stated goals, plans for systematic technology development, and stable financial commitment.
- Continue to follow developments in fire safety in the materials and aerospace industries, as well as in related industries. Coordinate within the U.S. Department of Transportation and with other federal agencies conducting related research, including the National Aeronautics and Space Administration, the departments of Defense, Energy, Transportation, Commerce, and the National Science Foundation.

REPORT ORGANIZATION

The findings of the committee have been organized into five chapters, with relevant background information included in the appendices. Chapter 1 introduces the study task and report objectives. Chapter 2 describes the array of design criteria that influence the selection and use of aircraft interior materials. Chapter 3 focuses on the evaluation and prediction of how aircraft materials perform in fires. Chapter 4 addresses the goal of developing substantially improved fire-resistant materials. Chapter 5 presents the committee's conclusions and recommendations for long-term research in fire-resistant materials development.

Appendix A is a glossary of fire-related terms. Appendix B provides a description of FAA research and developmental mandates in aviation safety and fire-research program plans. Appendix C discusses current fire-modeling capabilities. Appendix D provides a discussion of toxicity models and testing methods, and Appendix E contains biographical sketches of committee members.

1

Introduction

STATEMENT OF THE PROBLEM

The Aviation Safety Research Act of 1988 (P. L. 100-591) was aimed at improving air travel safety in the United States. The details of this legislation and related legislation (Federal Aviation Act of 1958, Aircraft Catastrophic Failure Prevention Research Act of 1990) are summarized in Appendix B.

The Aviation Safety Research Act of 1988 directed the Federal Aviation Administration (FAA) to spend at least 15 percent of its research budget on long-term investigations concerned with:

- aviation maintenance (addresses the aging fleet),
- fire safety (addresses fire containment and the fire resistance of engine fuel and cabin materials),
- human factors (addresses performance of flight crew, aircraft mechanics, and air traffic controllers), and
- dynamic simulation modeling of the air traffic control system (addresses air traffic capacity and control).

Following the Act's directive that, as part of fire-safety research, "The [FAA] Administrator shall undertake or supervise research to develop technologies and to conduct data analyses...to assess the fire and smoke resistance of aircraft materials..." the FAA Technical Center established a research program in advanced fire-safe materials and enumerated its goals in a high-level materials research and development plan (Lyon and Eklund, 1993), and also created a preliminary detailed plan for the work (Lyon, 1994). The FAA plans and goals for fire-resistant materials research are summarized in Appendix B.

According to the program plan, the objective of the program is to

> discover the fundamental relationships between the composition and structure of materials and their behavior in fires to enable the design of a totally fire-resistant cabin for future commercial aircraft. Research will be basic in nature and will focus on synthesis, characterization, modeling, and processing of new materials and materials combinations to improve the fire performance, increase the functionality, and reduce the cost of next-generation cabin materials.

The FAA Technical Center began acquiring staff and laboratory equipment for this program in 1993. The program is envisioned to be a long-range effort to develop fire-safe materials for use on future commercial aircraft that would represent an "order-of-magnitude" improvement in aircraft cabin fire-worthiness. The program has the following goals:

- determine the fundamental relationships between the composition and structure of materials and their behavior in fires,
- use this knowledge to identify and design new materials and material combinations that provide an order-of-magnitude improvement in fire-worthiness, and
- develop the processing technology to ensure manufacturability and recyclability of advanced fire-safe materials.

To provide guidance for this effort, the FAA requested that the National Research Council (NRC), through its National Materials Advisory Board, establish a committee to identify promising fire-resistant[1] materials technologies, component design issues, and performance parameters and to recommend research in promising areas. To carry out its charge, the NRC committee held a technical conference at which the participants assessed the state of the an of fire-resistant materials, reviewed ongoing research in improved materials, summarized significant findings, and suggested objectives for the FAA advanced fire-safe materials research program. Included in the published conference proceedings (NRC, 1995) are the conference workshop session summaries upon which the committee built its evaluation to arrive at its final conclusions and recommendations for this report.

ACCIDENT STATISTICS

In the development of their research initiatives, the FAA based much of its planning on how to counter the kinds of accidents that have occurred in the past. Aircraft accident

[1] *Fire resistance* is defined as the property of a materials or assemblage to withstand fire or give protection from it (ASTM, 1994).

statistics for commercial transport aircraft[2] and the of aircraft fires are summarized in this section.

The FAA reported that for United States transport airlines, its 1981–1990 database showed there had been 1,153 fatalities, of which 535, or approximately half, had been associated with nonsurvivable accidents, and that the National Transportation Safety Board 1964–1988 database showed that about one-third of the fatalities had been associated with nonsurvivable accidents (FAA, 1991). The FAA data show that about 60 percent of the fatalities in survivable accidents are due to impact trauma (i.e., 30 percent of total fatalities), and the other 40 percent are due to fire (20 percent of total fatalities).

This section presents a summary of accident data from 1959 through 1993, using the Boeing database, which is representative of all data (Boeing Commercial Airplane Group, 1994). These data cover Western-manufactured commercial transport aircraft heavier than 60,000 pounds gross weight. Turboprop aircraft are not covered. Aircraft manufactured in the former USSR are not included because that database is incomplete. Military operators of commercial-type aircraft are also not included.

As shown in Figure 1-1, the number of commercial aircraft has steadily grown from about 1,000 in 1964 to 11,433 in 1993. The number of departures (flights) has increased from less than 2 million in 1964 to 13.86 million in 1993. Figure 1-2 shows that the number of accidents per million departures decreased rapidly from the introduction of jet aircraft in 1959, and has remained relatively constant for the past two decades at about two accidents per million departures in worldwide scheduled passenger operations. Figure 1-3 shows that the number of fatal accidents per million departures also decreased rapidly after 1959 and has remained relatively constant for the past two decades at about one accident per million departures, with about 500 annual fatalities involving occupants of the aircraft. The number of non-fire-initiated accidents that involve fire is about 0.7 accidents per million departures. The number of fire-initiated accidents is about 0.1 accidents per million departures (Murray, 1995).

Table 1-1 shows a synopsis of accidents occurring in passenger and cargo, and test, training, demonstration, and positioning operations since jet aircraft were introduced. From 1959 through 1993, there were 398 fatal accidents with 19,298 fatalities, of which 319 accidents and 18,956 fatalities were in passenger aircraft. In the 10-year period 1984–1993 there were 120 fatal accidents with 5,526 fatalities, of which 96 accidents and 5,397 fatalities were in passenger aircraft.

About half of accidents occur during final approach and landing, which is only about 4 percent of the flight time. Figure 1-4 shows the primary causes for accidents that occur on final approach and landing.

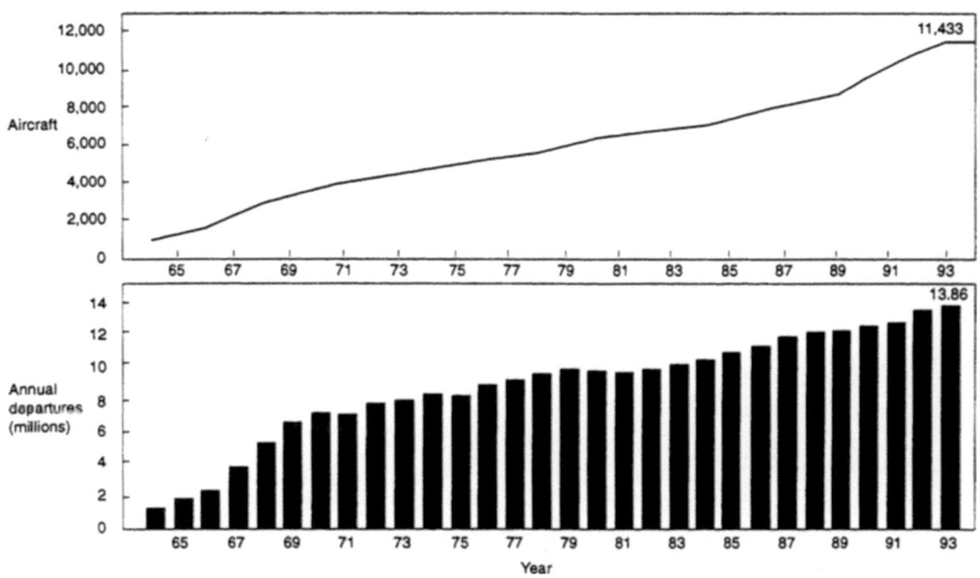

Figure 1-1 Fatal accident rates and fatalities. Source: Boeing Commercial Airplane Group (1994). Reproduced courtesy of The Boeing Company.

[2] This study concerns commercial transport aircraft, that is, certified jet aircraft greater than 60,000 pounds maximum gross weight including those in temporary nonflying status and those in use by nonairline operators, but excluding military (and former Soviet Union) operations (Boeing Commercial Airplane Group, 1994:3).

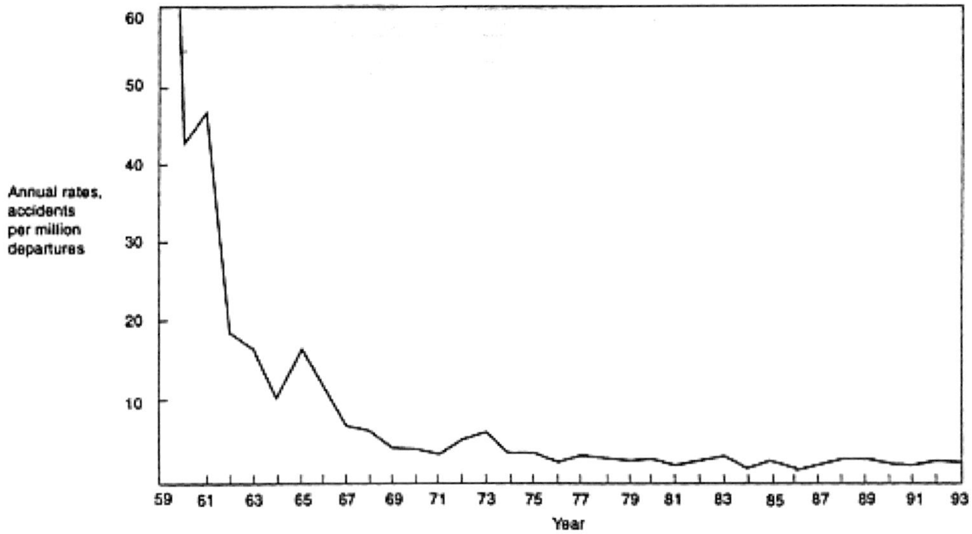

Figure 1-2 Jet aircraft in service and annual departure. Source: Boeing Commercial Airplane Group (1994). Reproduced courtesy of The Boeing Company.

Figure 1-5 shows the classification of accident types involved in airline fatalities. The predominant scenario has been controlled flight into terrain (CFIT), wherein the aircraft crashed into the ground while under control of the flight crew. CFIT accidents have been predominantly nonsurvivable.

Figure 1-6 shows the primary factors involved in fatal accidents. The data show that in those fatal accidents for which a primary cause has been identified (about 84 percent), the largest factor is attributed to flight crew (more than 50 percent), followed by airplane factors (less than 10 percent),

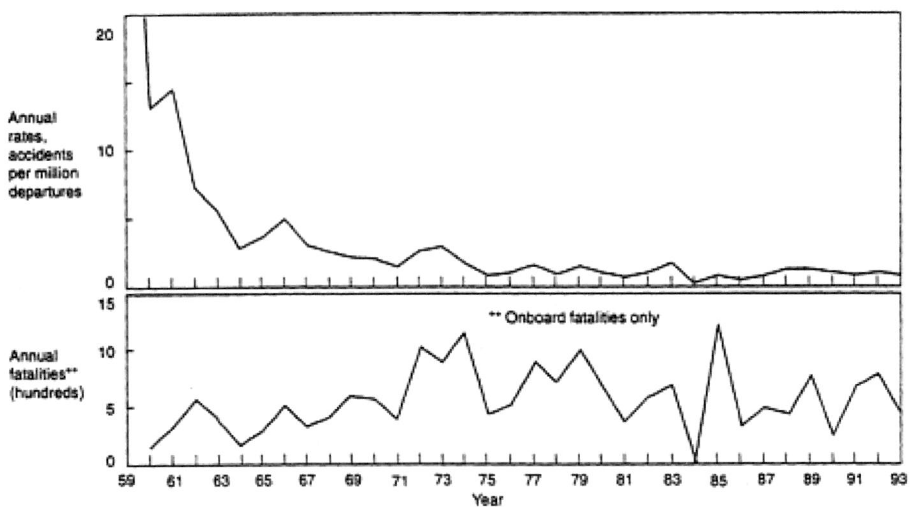

Figure 1-3 Annual accident rates. Source: Boeing Commercial Airplane Group (1994). Reproduced courtesy of The Boeing Company.

TABLE 1-1 Summary of Fatal Accidents

	Number of Fatal Accidents		Onboard Fatalities	
Type of operation	1959–1993	1984–1993	1959–1993	1984–1993
Passenger	319	96	18,956	5,397
All-cargo	44	18	174	76
Test, training, demonstration and positioning	35	6	168	53
Totals	398	120	19,298	5,526

Source: Murray (1995). Reproduced courtesy of The Boeing Company.

airport or air traffic control factors (less than 10 percent), and maintenance, weather, and miscellaneous items.

STATEMENT OF OBJECTIVES

There are two principal objectives of this study:

- Identify promising materials technologies, design issues (both overall and for individual components), and fire performance parameters (for both full-scale and individual components) that, if properly optimized, would lead to improved fire and smoke resistance of materials and components used in aircraft interiors.
- Identify those fundamental, long-range research topics that hold the most promise for producing predictive technology, new advanced materials, and the required product development to achieve totally fire-resistant and fire-safe interiors in future aircraft.

Aircraft are complex systems composed of many components of different materials. The scope of this study is limited to those materials and parts that compose the aircraft interior and are discussed more fully in Chapter 2. The aircraft interior is considered here to be any parts and materials within the

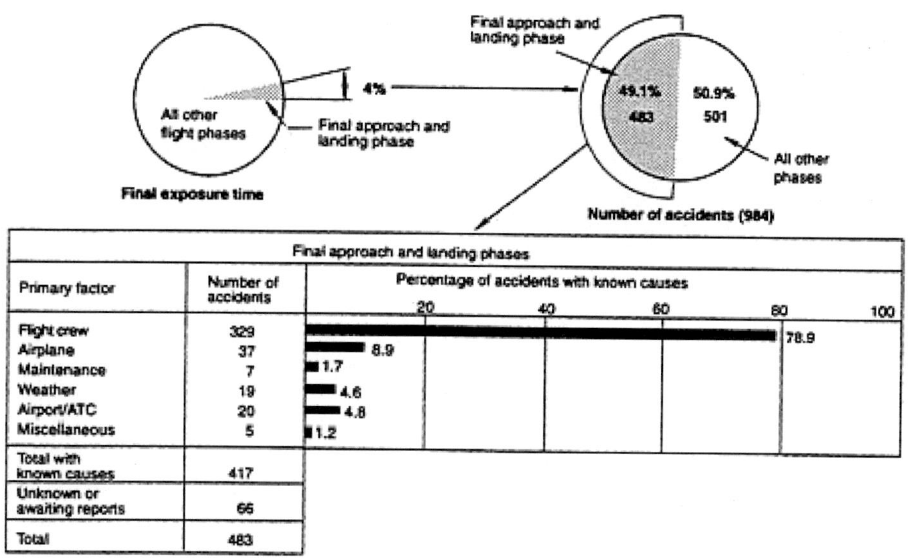

Figure 1-4 Causes of accidents in final approach and landing. Source: Boeing Commercial Airplane Group (1994). Reproduced courtesy of The Boeing Company.

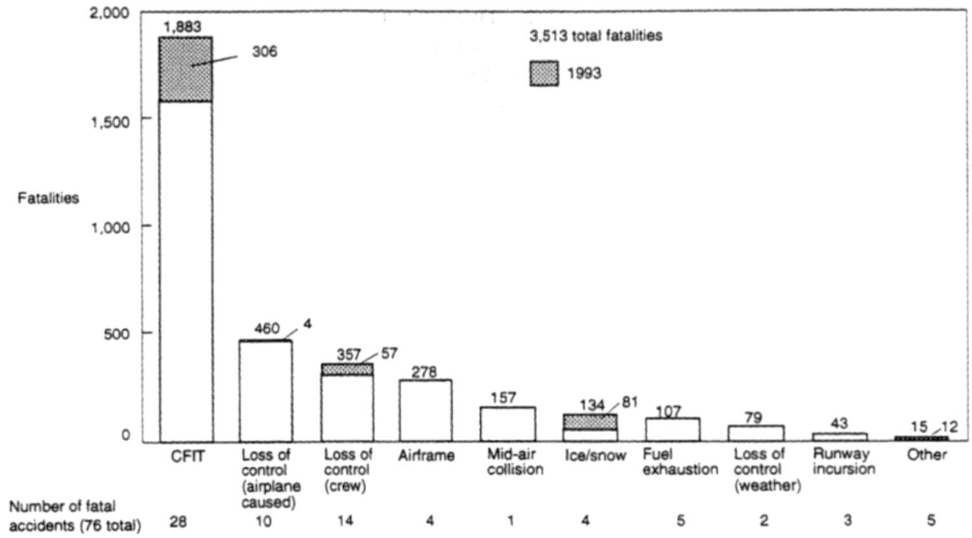

Figure 1-5 Fatalities classified by type of accident. Source: Boeing Commercial Airplane Group (1994). Reproduced courtesy of The Boeing Company.

Primary factor	Number of accidents: Total	Number of accidents: Last 10 years
Flight crew	217	50
Airplane **	33	8
Maintenance	11	5
Weather	14	1
Airport/ATC	24	7
Miscellaneous/other	37	23
Total with known causes	336	94
Unknown or awaiting reports	62	26
Total	398	120

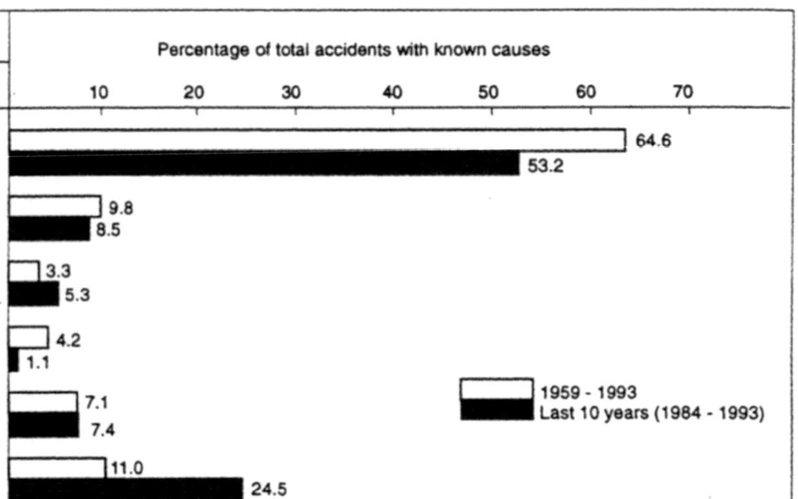

Figure 1-6 Primary causes of fatal accidents. Source: Boeing Commercial Airplane Group (1994). Reproduced courtesy of The Boeing Company.

structural shell of the aircraft, including cabin furnishings; ceiling, wall, and floor panels; cargo compartment and containers; insulation; wiring; lighting; ducts; windows; and lavatories. Payload items, such as luggage, items, other cargo, newspapers, magazines, the clothing of passengers, and food-related items are important to such fire-related issues as ignitability and generation of smoke and toxic products, but the fire resistance of these items is not considered in this study because they would be very difficult to control. However, the impact of such items as sources of heat in aircraft interiors during crashes should be assessed when evaluating and modeling the fire-resistance qualification and testing conditions. Fire safety systems such as water mist, hoods, vents, and fire suppression systems may affect the criteria used to establish performance goals for the development of fire-resistant materials. These systems, while providing additional benefits in fire safety, were not within the scope of this study.

Aircraft interior materials must be lightweight and meet engineering, wear, and cosmetic requirements, in addition to having desirable fire-safety characteristics. In meeting all of these requirements, it is important to recognize that there are not likely to be optimal materials that will, under all possible circumstances, be completely nonflammable and incapable of generating smoke and toxic products. Furthermore, materials processing requirements may limit the application of materials technologies that are otherwise preferred in terms of fire safety. These considerations dictate an approach to the principal study objectives set forth above that begins with defining the most likely fire scenarios to be experienced and how long the aircraft interior must remain safe once a fire occurs.

Although there have been some in-flight fires, catastrophic fires are generally post-crash related. In the past, in-flight fires have caused only a very small fraction of fire deaths. For example, according to FAA data, there has been only one in-flight fire death in U.S.-registered commercial transport aircraft (and that death was a suicide). In non-U.S.-registered aircraft, the few catastrophic in-flight fires were initiated in inaccessible areas.

The committee considered both in-flight and crash fire categories. However, in-flight fires in accessible areas within the aircraft interior were not considered. These fires have traditionally been extinguished quickly by properly trained flight personnel aided by early detection and warning systems and suitable portable extinguishing equipment. In-flight fires in inaccessible areas are more problematic because of the potentially long periods before passengers can be evacuated and the fire extinguished. Long-term fireworthiness and materials that produce extremely low levels of smoke and toxic products are required for such inaccessible aircraft interior components. In-flight fires in inaccessible areas were therefore one focus of this study.

For post-crash fire scenarios, the imperative is to provide passengers who survive the crash sufficient time to leave the aircraft without fatal exposure to heat and smoke and toxic fire products. Smoke and toxic products can result in visual obscuration and partial mental and physical impairment, thus indirectly increasing the required evacuation time. Post-crash fire scenarios are extremely varied; however, the committee defined several general categories for such fires in order to better classify typical fire exposures. Both post-crash and in-flight fire scenarios are described in Chapter 3.

Finally, this study emphasizes long-term innovation leading to assessments of the fireworthiness of aircraft interiors 10–20 years hence (see Chapter 4). Thus this report addresses the issue of aircraft interiors more than 10 years in the future, including possible high-speed civil transports and large subsonic transports.

2

Design and Function Requirements for Aircraft Interior Materials

For the purposes of this study, the interior of an aircraft is considered to be everything that is contained inside the pressure shell, that is, the pressurized part of the aircraft fuselage. The regulatory requirements of Federal Aviation Requirements (FAR), Part 25, that apply to interiors fall in FAR 25.853, Compartment Interiors, and FAR 25.855, Cargo or Baggage Compartments. Although FAR 25.853 has the introductory terms "For each compartment occupied by the crew or passengers, the following applies," items that ate not strictly in the occupied compartment (i.e., are outside the cabin liners and not visible to either crew or passengers, such as "electrical conduit, thermal and acoustical insulation and insulation covering, air ducting," are specifically cited and the regulatory requirements are also applied to them. Interior cabin liner components are identified in Figure 2-1.

An aircraft interior is designed to meet the requirements of:

- the FAA and other regulatory agencies,
- the airlines,
- airline passengers, crew, and
- the aircraft manufacturers.

There are minimum requirements that emanate from these four groups, which together comprise the design criteria. Issues to consider in the development of combined (inclusive) design criteria for aircraft interiors are shown schematically in Figure 2-2.

The safety criteria include the FAA regulatory mandates, which address only safety and axe largely quantitative. However, there are other, nonregulatory requirements such as passenger comfort level that are difficult to quantify, which complicates the task of the designers. Aircraft interior design is further complicated by the fact that many of these needs compete with each other and thus trade-offs are necessary.

Once the design of a part has been established by design engineering organizations, and once drawings describing the design and manufacture are released to the manufacturing organizations, many business processes are activated to car out acquisition of materials (inventory), tools, facilities, and manpower. If a subsequent change is made to the design, all the manufacturing planning is also subject to change, which can be time consuming and costly and creates the potential for a substantial economic penalty. There is, therefore, a strong priority assigned to designing parts "right the first time."

MATERIALS OF CONSTRUCTION

The current state of the art for materials used to make parts that satisfy the design criteria and other requirements fall into several main categories or families. Materials categories that could be used to fabricate more fire-resistant interiors would be subject to the same selection and use criteria.

Currently, most of the vertical and ceiling surfaces of aircraft are comprised of sandwich panels fabricated from face sheets of phenolic resin and fiberglass or carbon fiber reinforcement, and a polyaramid (Nomex®) core. These panels are covered with highly formable decorative thermoplastic films that are printed in a variety of complex patterns and

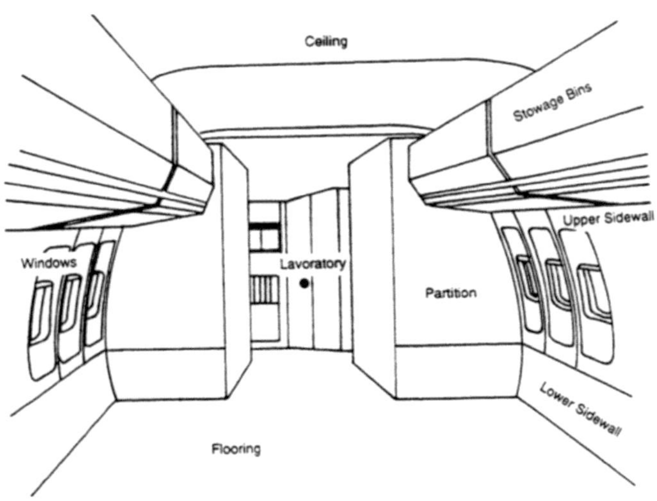

Figure 2-1 Aircraft interior cabin liner components. Source: FAA (1990).

Figure 2-2 Considerations in inclusive design criteria.

colors and are embossed in a wide selection of textures and gloss levels. These films are designed to meet rigid airline and manufacturing demands involving color, gloss, and texture. Although these characteristics are largely subjective and therefore difficult to specify and measure, they must nonetheless be consistently matched. Typical panel constructions are shown schematically in Figure 2-3.

Fire-resistant textiles have presented some especially difficult problems. Aesthetics demand that upholsteries, draperies, carpets, and tapestries be available in a wide variety of colors and have tight tolerances for look, feel, and durability. New-generation synthetic fibers with improved fire resistance have had a natural dark color that has essentially precluded their being pigmented in light colors, so the lack of availability of a wide variety of colors has greatly limited their acceptance for aircraft application (Hasselbrack, 1995).

The major material that is used for upholstery and drapery has been fire-retarded wool, with some use also made of a fire-retarded polyester, both of which meet the fire-resistance requirements and can also be dyed in an unlimited range of colors. Tapestries are held to more stringent flammability requirements than upholstery and drapery. It has been difficult to formulate a fire-retardant scheme for wool that allows it to meet the more-stringent requirements. Therefore, tapestries currently have to be fabricated of the new synthetic materials, with a fairly limited color palette or of wool/synthetic hybrid fabrics. This restriction has, to some extent, discouraged the use of tapestries. Other decorative schemes of less aesthetic appeal are being used in place of tapestries.

There are numerous other material types used in various applications. Examples are summarized in Table 2-1.

SAFETY

The air transport industry and its regulators have achieved an outstanding safety record (as described in Appendix B) by placing an intense, vigorous, and unrelenting priority on the safety of the air transportation system. To maintain and even improve this excellent aviation safety record, the people responsible for the operation and maintenance of aircraft—flight crews, airplane mechanics, and air traffic controllers—are selected and trained according to rigorous safety criteria.

Aircraft are designed to operate routinely under extreme conditions, that is, both at altitudes where human life cannot be sustained outside the aircraft and on the ground while taking off and landing at high speeds. The amount of fuel necessary to move such large and highly engineered machines over several thousand miles while at an altitude of seven or eight miles is tremendous. For example, the heat energy contained in the fuel carried by a 747 (more than 50,000 gallons of jet fuel) is more than 20 times as much as would be needed to heat up and melt the entire airframe.

Minimum safety standards for aircraft design, manufacture, and operation are established by FAA regulations. In addition to the regulatory mandates, aircraft manufacturers use supplemental design criteria that go beyond the regulatory requirements (for example, see Boeing Commercial Airplane Group, 1977; Airbus Industries, 1979).

The criteria for interior safety requirements were developed for normal operation (which includes all non-crash-related incidents) and for several survivable crash-related scenarios (crashworthiness). Although there are many types of criteria, the major ones are:

- structural strength and stiffness,
- fire resistance (includes control of smoke generation),
- interior configuration and emergency evacuation, and
- emergency oxygen systems.

Of these, fire resistance structural strength and stiffness have the most impact on research for improved fire- and smoke-resistant materials.

Fire Resistance

Although there are regulations concerning physical and mechanical properties as well as configuration and layout requirements, the FAA regulatory requirements for interior furnishings are based, in large part, on flammability. The flammability mandates for transport aircraft are listed in FAR 25.853, FAR 25.855, and FAR 25.869. For most furnishings (except cabin liners, seats, and cargo liners) these comprise Bunsen burner tests to characterize resistance to ignition and ability to sustain a flame. In addition to ignitability requirements, cabin liners are subject to additional requirements that

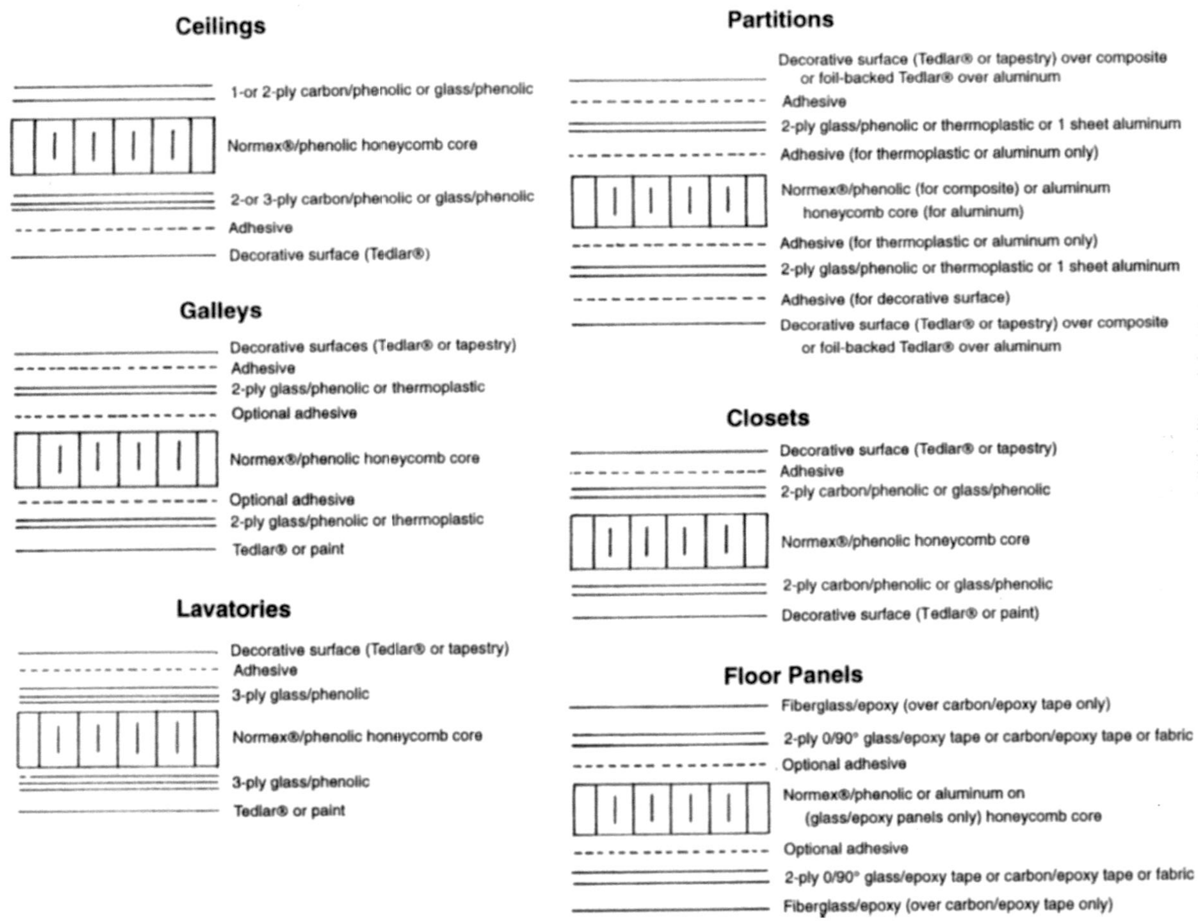

Figure 2-3 Typical cabin panel constructions. Provided courtesy of The Boeing Company.

TABLE 2-1 Materials Applications in Current Commercial Aircraft

Applications	Materials
Floor and floor covering	Glass or carbon/epoxy or phenolic/Nomex honeycomb floor panels
	-flexible urethane seat track covers
	-urethane foam edge band
	Mylar film over galley and entry floor panels
	Wool or nylon carpet
	-double-backed tapes to attach carpet to floor
	-Nomex felt underlay (at customer request)
	Poly(vinyl chloride) galley mats
Lower sidewall panel	Glass or carbon/phenolic/Nomex honeycomb plus scuff resistant surface (wool or Nomex fabric, or tough plastic)
Upper sidewall panel	Glass or carbon/phenolic/decorative thermoplastic layer plus Tedlar
Light covers	Polycarbonate
Overhead stowage bins	Glass or carbon/phenolic/Nomex honeycomb plus edge urethane foam layer plus reinforcement
Gap fillers	Silicone or urethane
Passenger and cabin attendant seats	Wool, wool/nylon, or leather upholstery
	Urethane foam cushions
	Polybenzimidazole or Nomex/Kevlar blocking layer
	Polyethylene form flotation foam
	Thermoplastic seat trays
Partitions	Glass or carbon/phenolic/Nomex honeycomb
	Decorative thermoplastic laminate or wool/Nomex textile or leather
	Polycarbonate transparent wind screen (infrequent)
Stowage bins	Glass or carbon/phenolic/Nomex honeycomb
	Decorative thermoplastic laminate
	Wool textile interior liner (infrequent)
Placards	Poly(vinyl chloride) or urethane
Insulation	Fiberglass batt, phenolic binder, Mylar cover
	Poly(vinyl chloride)/nitrile rubber, polyethylene, foams
	Polyimide foam
Windows	Outer pane stretched acrylic
	Inner pane cast acrylic
	Dust cover polycarbonate or acrylic
Passenger service units	Molded thermoplastics (Ultem, Radel, polyethelketoneketone)
	Aluminum
	Glass or carbon/phenolic
Hoses	Silicone
	Nylon
	Urethane
Air ducts	Glass/phenolic, epoxy, or polyester for large ducts
	Polyisocyanurate foam for large ducts
	Fire-retarded nylon
	Glass/silicone
	Nomex felt (small quantity)
	Polyimide foam wrap

Source: NRC (1995).

TABLE 2-2 FAA Flammability Requirements for Cabin Liners (sidewalls, ceilings, and partitions)

Test Type	Current Minimum Acceptance Criteria
Ignitability (60-second vertical Bunsen burner)	6-inch bum length[a] 15-second specimen extinguishing time[b] 3-second drip extinguishing time[c]
Heat release (Ohio State University calorimeter)	65 kW/m^2 peak rate (during a 4-minute test).[d] 65 kW • min/m^2 total (during the first 2 minutes)[e]
Smoke release (National Bureau of Standards smoke chamber)	200 specific optical density (during a 4-minute test)[f]

NOTE: Definitions and test procedures are described in detail in FAA (1990).

[a] *Burn length*: the distance from the original specimen edge to the farthest evidence of damage to the test specimen due to that area's combustion.

[b] *Specimen extinguishing time*: the time that the specimen continues to flame after the burner flame is removed from beneath the specimen.

[c] *Drip extinguishing time*: the time that any flaming material continues to flame after falling from the specimen to the floor of the test chamber.

[d] *Heat release rate*: the rate at which heat energy is evolved by a material when burned. The maximum heat release rate occurs when the material is burning most intensely.

[e] *Heat release*: a measure of the amount of energy evolved by a material when burned.

[f] *Specific optical density*: a dimensionless measure of the amount of smoke produced per unit area based on light transmittance measurements.

involve control of total heat release and heat release rate and density of smoke produced. Seats and cargo liners must meet rather severe tests based on kerosene oil burners of the sort used in home heating furnaces. Detailed descriptions of flammability test methods for individual aircraft components are described in "The Materials Fire Testing Handbook" (FAA, 1990). Table 2-2 summarizes the flammability requirements for cabin liners.

Current regulations covering the flammability of interior cabin furnishings apply directly to individual parts that make up the interior, for example, those comprising sidewalls, ceiling, partitions, stowage bins, windows, air ducts, and insulation. There is a multitude of such parts; for example, sidewalls consist not just of multiple copies of certain parts, but of a large number of parts with different part numbers. Other components also make up long lists of part numbers.

Many of the current fire-related regulations were based largely on recommendations of the FAA SAFER committee (FAA, 1980). The SAFER committee focused on ways to make post-crash fires more survivable. Their findings concerned fire hazards associated with spilled fuel ignition and burning, hull burnthrough, and involvement of cabin materials and escape slides. The recommendations to the FAA that were associated with fire-resistant materials included:

- establishing contribution of cabin materials to fire hazard based on large-scale tests;
- developing seat fire-blocking layers;
- expediting the development and evaluation of the Ohio State University (OSU) heat-release test chamber;
- accelerating toxicity research;
- amending the flammability test methods to account for melting and drip behavior of certain materials;
- defining fire scenarios for modeling, research, and design; validating models with small-and full-scale tests;
- establishing radiant heat-resistance standards for inflatable evacuation devices;
- expediting the development of fire-resistant cabin windows; and
- supporting development of fire-resistant cabin interior materials and encourage the development of a materials data bank.

Action based on many of the SAFER committee recommendations have contributed to the development of current flammability standards, including seat fire-blocking regulations, cabin liner materials heat release regulations using the OSU test, and a Technical Standard Order requiring radiant heat resistance for escape slides.

Strength and Stiffness Requirements

The design criteria include strength and stiffness requirements. All cabin interior components must be designed to routinely withstand "limit loads" (i.e., typical flight loads).

For a margin of safety, components must also be able to withstand occasional "ultimate loads" (1.5 times limit loads) and "assist loads" or "abuse loads" (e.g., bumping, pushing, and pulling handles). Components are designed to undergo only elastic deflections under these conditions.

In addition to the loads cited above, some components such as seats, stowage bins, closets, and class dividers must also be able to safely restrain items of mass (e.g., passengers or stowed items) under "minor" crash loads, as specified in FAR 25.561:

(a) The airplane, although it may be damaged in emergency landing conditions on land or water, must be designed as prescribed in this section to protect each occupant under those conditions.

(b) The structure must be designed to give each occupant every reasonable chance of escaping serious injury in a minor crash landing when—

(1) Proper use is made of seat belts and other safety design provisions;

(2) The wheels are retracted (where applicable); and

(3) The occupant experiences the following inertia forces acting separately relative to the surrounding structure:

(i) Upward, 3.0g

(ii) Forward, 9.0g

(iii) Sideward, 3.0g on the airframe; 4.0g on the seats and their attachments

(iv) Downward, 6.0g

(v) Rearward, 1.5g

(c) The supporting structure must be designed to restrain, under all loads up to those specified in paragraph (b) (3) of this section, each item of mass that could injure an occupant if it came loose in a minor crash landing.

(d) Seats and items of mass (and their supporting structure) must not deform under any loads up to those specified in paragraph (b) (3) of this section in any manner that would impede subsequent rapid evacuation of occupants.

Furthermore, the seats and seat/seat track attach points must meet dynamic load conditions that underlie criteria for restraint loads and head injury, leg injury, and seat deformation.

From the above mandates on the components themselves, the designer must evaluate the loads the constituent materials must sustain and what the measurable physical and mechanical properties need to be. Typical mechanical properties include tensile strength and stiffness; compression strength and stiffness; shear strength and stiffness; flexural strength and stiffness; impact strength; and pull-out, torque, and shear strengths of insert installations.

To establish mechanical property limits used in design, materials (especially those used in parts that must bear structural loads) are thoroughly tested at various representative temperatures and environments that the material might experience during its service lifetime. Beyond this, however, the certification of components that would potentially bear critical loads requires that each newly designed component be structurally substantiated by comparison for similarity to previously certified components, analysis, or test.

Components are designed to absorb crash loads through component deformation (yield). For example, controlled deformation of seat structure is the primary means through which designers have been able to meet the requirements for dynamic loading of seats and seat track attachments. If the seat were totally rigid and all the crash energy transmitted directly to the passenger, the passenger may not survive the crushing effect of a safety belt.

WEIGHT

Weight is one of the most important criteria in designing an aircraft. The economic viability of an aircraft's operation puts a priority on cost, payload, and range, which are directly impacted by weight. Airplane designers are keenly aware of the value of a pound of weight, which is a cost derived from the fact that if the empty airframe weight is reduced by a pound, not only is less fuel needed for the airplane's operation for a given range, but also more payload and fuel (more fuel equals more range) can be carried. Over the life span of an airplane in service (typically more than 20 years), the additional fuel needed to transport an extra pound on the airframe amounts to $200–$400 at current fuel prices ($0.60–$0.80 per gallon at 6.66 pounds per gallon) .[1] The cost of losing payload or range is more difficult to establish, but nevertheless, it puts the economic viability of the aircraft at risk.

COST

Cost is another critical criterion in designing an aircraft. The cost of procuring and maintaining an aircraft interior

[1] An average wide-body aircraft on a 3,000-mile flight might expend 0.25 pounds of fuel for every additional pound to the interior. Considering 10 such flights per week for a year and current fuel costs ($0.60 per gallon), the aircraft's operating cost would increase by $12 per year per additional pound. For a narrow-body jet, assuming 1,200-mile flights and 24 flights per week, the additional fuel cost would increase by $20 per year per additional pound.

factors into the total cost of running an airline and must be taken into account in interior design and manufacture. Technological advances must "earn" their way into implementation by means of either resultant decreases in operating costs or customer-approved improvements in operation that justify increased costs.

Since all airlines, for marketing reasons, strive to present a specific and identifiable image, the interior furnishings tend to have a unique decor and are therefore in a sense custom made, which results in higher costs. The costs of such detailed design and manufacture must therefore be amortized over fewer units of aircraft than many mass-produced commodities. There is a priority, therefore, to optimize the efficiency of all processes involved to provide cost reduction.

The manufacturing costs for components are composed of materials costs and processing costs. Reducing each of these has a priority, but the installed part cost is decisive. In this way, an expensive material that is easily and inexpensively processed may be favored over a less expensive material that is difficult to process.[2]

MANUFACTURABILITY AND PRODUCIBILITY

The need to reduce production costs has put great pressures on aircraft designers, including cabin interior designers, to create designs with a priority on the ease of manufacturing and assembly.

One method that has produced good results involves teams that include representatives from engineering, quality assurance, and manufacturing working together to create designs that can be manufactured more cost effectively. A component's form, fit, function, quantity required, and other design criteria are combined with the various possibilities of materials and processes (namely, materials cost, processing labor, and capital equipment and tools) to define the final part.

Standard parts that are interchangeable among all customers (e.g., air-conditioning ducts and cargo liners) need to be designed with materials and processes that minimize processing labor since they can be produced in higher volumes. Customized parts, such as cabin liners that have a common construction for all customer airlines but a customized texture and decor, need to be designed with materials and processes that allow rapid change and flexibility.

Processing Labor

As a general rule, processing labor accounts for most of the cost of a pan. However, there is a widespread range of processing costs that depend on the materials and manufacturing processes selected.

Factors that must be considered when designing nonmetallic parts include:

- part quantity;
- capability for integration of multiple details;
- texturability;
- control of surface smoothness;
- thickness consistency;
- edge close-outs;
- bonding;
- machining and drilling;
- trim and finish;
- conditioning;
- environment and temperature compatibility when coforming dissimilar materials;
- storage, handling, and shelf life;
- compatibility with fasteners and inserts;
- part size; and
- interchangeability.

Examples of processes commonly available for the manufacture of aircraft interior parts include composite lamination and press curing, compression molding, hydroforming, casting, pultrusion, and filament winding of thermosetting polymers and composites and blow molding, rotational molding, thermoforming, injection molding, casting, and extrusion molding of thermoplastic polymers and composites. The most common processes used in the fabrication of components for current aircraft cabins are composite lamination followed by press curing. Honeycomb core with fiber-reinforced phenolic face sheets make up the majority of cabin interior panels (Berg, 1995). Common thermoplastic processes currently used include sheet thermoforming and injection molding.

The amount of processing labor depends on the part configuration and the process used. For example, a part constructed from a fiber-reinforced composite laid up by hand takes considerably more time to prepare, lay up, cure, trim, and finish compared to the time required to process an injection-molded part. It is therefore important that the design of a part take account of the manufacturing process that has to be used to produce the design and that the materials and processes are jointly optimized to minimize manufacturing cost.

---

[2] In general, the procurement cost to the airline of a component, and the influence of materials and processing costs, is similar whether the component is part of the delivered equipment or installed subsequently. In the case of a retrofit or refurbishment, the added cost to the airline would be associated with processes such as removal and disposal of the old component, development of plans and procedures for maintenance crews to follow, airplane downtime, etc. These costs have more to do with regulatory mandates that may compel upgrades of the existing fleet than with materials development and selection decisions and priorities.

Processing Equipment and Tools

A part's design and manufacturing process affects the tooling required as well as the processing labor. During component design, designers need to ensure that the manufacturing tooling required is compatible with the part's size, form, and the manufacturing process. The quantity of parts to be produced also affects the material and process selection. For example, injection molding is appropriate only for production of large quantities of parts because injection-molding tooling is expensive.

COMFORT

Materials properties, component design, and the interior layout (configuration) all contribute to passengers' convenience and comfort. Important comfort factors for this study are seat ergonomics and acoustics.

Seat Ergonomics

Aircraft manufacturers do not design or produce passenger seats. Although aircraft manufacturers provide standard seat specifications to ensure correct fit and adherence to safety and certification requirement, passenger seats are manufactured by companies that specialize in them. Generally, airlines work directly with seat manufacturers to specify and purchase passenger seats. The seats are then manufactured to the airline's specifications, which must of course include provision for appropriate means for attachment to the seat tracks on the cabin floor that are provided by the aircraft manufacturer. Noisily, the seats are then shipped to the aircraft manufacturer for installation in the cabin prior to delivery of the airplane, although in some instances the seats are shipped tot he Customer airline, which takes delivery of the airplane and subsequently installs the seats at its own maintenance facilities. Typical seat construction is shown in Figure 2-4.

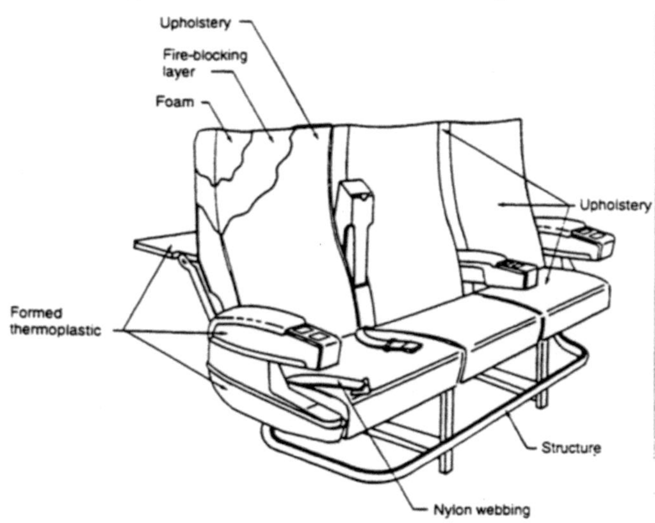

Figure 2-4 Typical seat construction. Source: FAA (1990).

In the 1970s it was recognized that the greatest fire load in the passenger cabin was by far the seat cushion foams which were exclusively fire-retarded flexible polyurethane. In 1986 the regulatory requirements were changed to require that the upholstery/cushion combination satisfy a severe fire-resistance test. This mandate essentially required that the seat cushions be changed from the traditional fire-retarded urethane foam. The first available solution was to encapsulate the urethane foam in a durable fire-blocking fabric. Subsequently very highly fire-retarded foams (basically filled urethanes) were developed that also met FAA requirements. Alternatives to urethane foams for seat cushions have not been successfully developed.

The upholstery used on seats is typically a highly durable woven fabric of wool or a wool/nylon blend (the amount of nylon is relatively small, with 10 percent being the usual amount). While a type of fire-retarded polyester and leather have also been used as seat upholstery, no other natural or synthetic fiber type has been successfully implemented because of limitations with respect to fire resistance, color, durability, availability, color fastness, and other criteria (Hasselbrack, 1995).

Acoustics

Interior acoustics comprise a significant comfort factor. The major portion of cabin noise is generated by the engines and by the aerodynamic boundary layer along the exterior surface of the fuselage. Noise is reduced through absorption and transmission-loss mechanisms. Higher frequencies (above 500 Hz) are more easily absorbed, whereas lower frequencies must be blocked with mass (transmission loss).

Acoustic insulation is provided by the cabin insulation blankets, which also serve as thermal insulation. The thermal/acoustical insulation is comprised primarily of spun fiberglass that has a binder to hold the strands together. Owing to the acoustic requirements, the fiber diameter is very much smaller than the fiber diameter of typical building insulation. Also, there have been foams, typically of polyimide, that have been used in recent years to supplement the fiberglass.

The insulation material is encased in a thin film to hold it in place. The film is typically polyethylene terephthalate (Mylar®), although in the past polyvinylfluoride (Tedlar®) and polyimide (Kapton®) have also been used. Kapton films have a substantial weight and cost penalty compared to Mylar. Insulation blankets, used in areas where low-frequency sound must be blocked (e.g., near engines), are encapsulated in a heavy rubberized sheet material, such as a zirconium-filled silicone rubber.

Cabin linings also help to absorb some of the higher frequencies, but more importantly block some of the lower frequencies. Once in the interior, high-frequency cabin noise is reduced by using porous cloth upholsteries, carpets, and draperies. The sound penetrates these materials and becomes absorbed. Sound waves more easily bounce off leather upholsteries. The system of the insulation blankets, cabin linings, and furnishings combine to reduce exterior-generated sound by as much as 80–130 decibels.

In acoustic design, there is a desirable limited level of background noise in the cabin to provide a measure of privacy. However, as the noise level rises above this background level, it at first becomes difficult to carry on a conversation with a person close by, and then the noise level becomes physically discomforting. Acoustic designers pay attention to the "speech interference level," that is, the 1,000-, 2,000-, and 4,000-Hz bands, to avoid undue interference with speech.

In the event of a post-crash fuel-fed fire, the fiberglass insulation (if it remains attached) also provides a degree of protection as a fire barrier to inhibit the spread of fire to the cabin interior. The polyimide foams can also serve as such a barrier.

MAINTAINABILITY

Maintenance activities represent a significant part of the time and expense related to the operation of commercial aircraft. A cost-effective maintenance program is critical to the implementation of new materials or component designs. Maintainability has to do with the ease of preserving the passenger cabin interior in a satisfactory and usable condition. For interior furnishings, maintainability depends on component cleanability, durability, and repairability.

Cleanability

The major part of maintainability is the ease with which furnishings can be cleaned of normal soiling. Normal soiling agents consist of many things that are typically found in ordinary airline operations. Those that are difficult to remove include some types of dirt and mud; food stains such as mustard, catsup, grape juice, and red wine; cosmetics such as lipstick; and tobacco tar from cigarette smoke. Traditionally, candidate interior furnishings have been tested for cleanability before being implemented (i.e., for how easily common staining agents can be removed). Materials that are highly resistant to abuse such as scuffing and scratching, to color degradation by agents such as ultraviolet or even visible light, and to staining by common staining agents are required.

In general, a thin film of polyvinylfluoride or equivalent has been found to be nearly optimum with respect to these in-service requirements. Less desirable are paints from which stains are more difficult to remove (repainting is often necessary) and which are more easily degraded.

Flooring in some areas such as galleys, entry ways, and lavatories must be water resistant and skid resistant and easily cleanable for sanitary as well as aesthetic reasons.

Durability

Durability is how well components wear over time in everyday airline operation. To be economically viable to an airline's operation, an airplane—including its interior—must be durable.

Durability factors and requirements important for aircraft interiors include:

- resistance to vibration,
- colorfastness to light,
- resistance to tearing,
- resistance to crocking (textiles),
- dimensional stability,
- fluid resistance (e.g., solvents and cleaning fluids),
- resistance to permanent staining,
- resistance to water and moisture absorption,
- resistance to water wicking,
- resistance to corrosion,
- resistance to fungus attack,
- resistance to abrasion damage,
- resistance to impact damage, and
- resistance to crazing.

Many of these factors are obvious. For example, cabin liners. class dividers, and other exposed surfaces need to be reasonably resistant to mild abuse, such as the inadvertent contact with ends of stick-shaped items such as umbrellas. Also, stowage bins need to be quite tolerant of, for example, attempts to force in items that are too large or passengers who cannot quite reach up to the bin hanging onto the bin bottom. It is not difficult to envision other examples.

Other factors have more subtle implications. Thermal/acoustical insulation and wiring materials, for example, must be resistant to vibration damage. Poor thermal/acoustical insulation batting may be subject to crumbling and settling which would reduce the insulation efficiency, and poor wire insulation may be prone to chafing which could increase the risk of electrical arcing and fire.

Because of the extensive use of aluminum in current aircraft, it is necessary to be concerned with galvanic corrosion. Thermal/acoustical insulation also needs to be resistant to wicking and water absorption to avoid promoting corrosion. Textiles can cause corrosion problems due to the fire-retardant treatments involving flame-inhibiting salts that may promote corrosion if brought into contact with metal surfaces. Fabric treatments need to be tested to ensure they do not promote corrosion. Corrosion resistance is also pertinent to some of the newer composites that use reinforcement fabrics of carbon and graphite, which have galvanic potentials considerably different from those of metals, especially aluminum. Special sealing requirements are required to electrically isolate these composite materials from aluminum components such as seat tracks, fasteners, and floor beams.

Repairability

Even though interior furnishings may be durable, there is inevitable damage that occurs which requires that the affected item either be repaired or replaced.

In general, it is substantially less costly to repair damage to an interior furnishing than to discard and replace it. Although a damaged furnishing may be removed and replaced to allow the airplane to continue in uninterrupted service, the airline normally pursues repair of the item and stores it as a spare. Therefore there is a priority for repairability of reasonable damage to interior furnishings and to have repair processes and procedures identified and documented by the manufacturer.

AESTHETICS

Airlines use an airplane's interior decor, and to some extent the exterior decor, as a marketing tool. This is intended to reflect the airline's corporate image, style, and mission. Aircraft interior designers therefore need to have a variety of colors, prints, and textures and different degrees of plushness for airlines to select from to build a unique image and identity.

For newly manufactured aircraft, the basic shape and architecture of the interior are normally the standard ones offered by the aircraft manufacturer. The choices the airline customer has are mainly centered on decorative schemes to reflect the corporate image and perhaps the placement of some amenities such as lavatories and galleys.

Interiors for newly manufactured aircraft are normally designed by the aircraft manufacturer in coordination with the customer airline, but their actual fabrication is done by various combinations of the aircraft manufacturers and subcontractors who specialize in the manufacture of interior furnishings.

With proper maintenance, interiors are designed to last as long as the aircraft. Airlines often update the interiors of aircraft in their fleet to present a "new look." Interiors may also be refurbished or remodeled at some time during service due to transfer of the aircraft between airlines or safety-related retrofit. These interior furnishings are for the most part also fabricated by manufacturers who specialize in the manufacture of interior furnishings.

In both newly manufactured and redecorated aircraft, different levels of decor and amenities are used to target different segments of the passenger population. Business and first class travelers expect and are usually provided a higher level of comfort and luxury than passengers in the economy class; however, owing to the competition between airlines in the pursuit of passengers, the interior furnishings for all classes of service receive considerable attention.

The types of routes an airline flies and the airports it serves also affect aesthetic considerations. For example, airlines that fly into airports serviced by airstairs have their cabin interiors substantially affected by weather. Hot tarmac particles in the summer, snow in the winter, and combinations of rain, wind, and dirt during all seasons are constant problems. Another consideration is that airlines that provide international service to different areas of the world must contend with aircraft serviceability and maintenance requirements at locations far removed from their home base.

3

Evaluation of Materials Fire Performance

The development of improved fire-resistant materials for aircraft interiors requires an understanding of aircraft fire scenarios, the factors that influence the initiation and propagation of a fire, and testing and modeling methods to evaluate and predict materials performance. This chapter provides an overview of the current state of fire-safety science as it pertains to aircraft fires. The fire scenarios that are most likely to occur are described along with the parameters that are critical in each scenario. Predictive models for simulating fires and assessing fire hazards are summarized. Finally, testing methods useful in the characterization of materials performance in a fire as well as in the evaluation and ranking of new materials are described.

FIRE SAFETY

When evaluating a material for use in a given situation, it is important to assess the total fire hazard. *Fire hazard* is "the potential for harm associated with fire" (ASTM, 1994). Fire hazards are associated with the environment and with a number of characteristics of materials, products, or assemblies; these characteristics include ease of ignition, flame spread, rate of heat release, smoke generation and obscuration, toxicity of combustion products, and ease of extinguishment.

Fire risk is "an estimation of expected fire loss that combines the potential for harm in various fire scenarios that can occur with the probabilities of occurrence of those scenarios .... Risk may be defined as the probability of having a certain type of fire, where the type of fire may be defined in whole or in part by the degree of potential harm associated with it, or as potential for harm weighted by associated probabilities. However it is defined, no risk scale implies a single value of acceptable risk. Different individuals presented with the same risk situation may have different criteria for determining its acceptability" (ASTM, 1994).

The factors that need to be considered in assessing fire risk for aircraft interiors are the quantity of material present; component configuration; proximity of other combustibles; volume of the compartments to which the combustion products may spread; ventilation conditions; ignition, combustion, and toxic potency properties of the materials present; presence of ignition sources; presence of fire protection systems; number of occupants; and the time available to escape (Levin et al., 1982).

Flammability

Flammability describes the ability of a material to burn. Flammability characteristics are those properties that define, describe, or measure the behavior of a material when it is exposed to heat or fire. The flammability characteristics that are most important in measuring the material burning process are ignitability, flame spread, rate of heat release, and smoke and fire-gas production. Each characteristic is described in the following paragraphs.

Ignitability, or ease of ignition, measures the time it takes for a material to ignite once heat is applied. Ignition is the initiation of combustion as shown by glow, flame, detonation, or explosion. There are two different types of ignition processes: piloted ignition and nonpiloted ignition. *Piloted ignition* is the initiation of combustion as a result of contact of a material or its vapors with an external, high-energy source such as a flame, spark, electrical arc, or glowing wire (ASTM, 1994). In piloted ignition the surface temperature must exceed a critical minimum value (the firepoint) at which the rate of pyrolysis in the surface layer produces a flow of flammable vapors sufficient to support a flame (Drysdale, 1995). In *nonpiloted ignition* , no localized, sufficiently high-temperature region occurs in the gas phase to cause ignition, thus the surface temperature of a material must become high enough to act as an induced pilot to initiate gas-phase oxidation reactions and attain ignition.

Flame spread is a factor that signifies the rate of burning and is derived from the rate of progress of the flame front. Flame spread is affected by many parameters including surface orientation, material thickness, surface roughness, ambient pressure, humidity, size, initial fuel temperature, incident radiation, and material chemical composition.

Rate of heat release measures the rate of heat energy produced by a given amount of burning material. Heat release rate is a measure of the contribution of a material to a fire. The rate of heat release is considered the single most important measure of the fire hazard of a material. It is a quantity that can be used in predicting the rate of fire growth and its effects.

Smoke and fire-gas production are determined by the chemical composition of the material, the fire environment, and in particular the available oxygen. Smoke is defined as the airborne solid and liquid particulates and gases evolved when a material undergoes pyrolysis or combustion (ASTM, 1994). Fire gases, according to ASTM (1994), are airborne products emitted by a material undergoing pyrolysis or combustion that exist in the gas phase at the relevant temperature. Smoke density is influenced by composition, the rate of burning or intensity of the fire and the degree of ventilation.

Fundamental flammability principles are as follows (NFPA, 1988):

- An oxidizing agent, a combustible material, and an ignition source are essential for combustion.
- The combustible material must be heated to its piloted ignition temperature before it will ignite or support flame spread.
- Subsequent burning of a combustible material is governed by the heat feedback from the flames to the pyrolyzing or vaporizing combustible.
- The burning will continue until the combustible material is consumed, or the oxidizing agent concentration is lowered to below the concentration necessary to support combustion, or sufficient heat is removed or prevented from reaching the combustible material to prevent further fuel pyrolysis, or the flames are chemically inhibited or sufficiently cooled to prevent further reaction.

Toxicity

The toxic gases and irritants that are present in all smoke should be considered potential dangers. Toxic products can cause both acute and delayed toxicological effects. It is the acute and extremely short-term effects that prevent escape from an aircraft fire by causing faulty judgment, incapacitation, and death. The irritants in the smoke can also interfere with the ability of passengers to escape by causing severe coughing and choking and by preventing them from keeping their eyes open long enough to find the exits. In addition, delayed effects, such as tissue or organ injury, mutagenicity, carcinogenicity, and teratogenicity, may ultimately lead to permanent disability and post-exposure deaths among accident survivors.

Toxic potency is "a quantitative expression relating concentration (of smoke or combustion gases) and exposure time to a particular degree of adverse physiological response (e.g., death on exposure of humans or animals).... The toxic potency of smoke from any material, product or assembly is related to the composition of that smoke which in turn is dependent upon the conditions under which the smoke is generated" (ASTM, 1994). The LC_{50} (the concentration that causes death in 50 percent of the test organisms in a specified time) is a common endpoint used to assess toxic potency. In the comparison of the toxic potencies of different compounds or materials, the lower the LC_{50} (i.e., the smaller the amount of material necessary to reach this endpoint), the more toxic the material is.

FIRE SCENARIOS

For fire-resistant materials, self-sustaining interactions of individual materials are of minimal importance to the real fire scenario, even in terms of local ignition and sustained smoldering.[1] Fire resistance is generally sufficiently effective to inhibit flammability under all but extreme heat loads resulting from either external source coupling or large-scale involvement. Thus, important issues are dominated by systems interactions to heat exposure from surroundings rather than the fire characteristics of individual materials themselves. However, the nonthermal hazards (visibility, production of irritant gases, and toxic product generation) are more strongly connected to the characteristics of individual materials in response to these surrounding interactions.

In general, the dominating initial heat source is that radiated or convected from surrounding fuel fires. In large-scale fires such as fuel pool fires, radiative heat transfer is expected to dominate the convective component (Hottel, 1959). In the case of aircraft interior fire characteristics, pool fire plume impingement on the inside of the cabin through open egresses or structural failures can also be important.

Once the fire spreads to the inside of the aircraft interior. the heat release from the burning of local materials can contribute to further evolution of the overall fire scenario. eventually leading to flashover. Understanding space flashover and predicting the likelihood of its occurrence is Critical in assessing materials response in aircraft fires. *Flashover* is the point at which most of the combustible

[1] *Smoldering* is the combustion of a solid without flame (ASTM, 1994) and is characterized by a glowing combustion supported by strong exothermic char oxidation reactions. Smoldering usually occurs in well-thermally-insulated, natural polymers such as cellulosic materials and wood products and rarely occurs in synthetic polymers. The cables used in transport aircraft are generally made of Kapton2® which generates a lot of char but it is not known to smolder. Phenolic composite and Tedlar (polyvinyl fluoride) constructions used commonly in interiors do not smolder nor do the fiberglass or polyimide foam materials used in cabin insulation. Smoldering is more often observed in buildings where cellulosic insulation or cotton fabrics are used in upholstered furniture.

materials in an enclosed space reach their ignition temperatures at essentially the same time, so that the materials seem to burst into flame simultaneously. Flashover is a significant concern because the space becomes untenable for occupants, and the fire hazard to adjacent spaces increases significantly. Flashover conditions may, however, be preceded by such severe aircraft interior temperatures and nonthermal hazards as to preclude survivability. The FAA determined in its work during the development of the heat release regulation that the toxicity hazard from burning standard cabin furnishings did not become significant until flashover occurred (Sarkos and Hill, 1989).

Catastrophic fire events can be the result of two basic aircraft fire scenarios—post-crash external fuel fires and in-flight fires (Sarkos and Hill, 1989). These two scenarios provide the basis for establishing baseline fire performance behavior and criteria for new materials development.

Post-Crash External Fuel Fires

The FAA Technical Center has developed criteria for the current generation of improved fire-resistant cabin materials based on the characterization of the fire environment through a series of full-scale tests (Sarkos, 1995). The scenario that has been emphasized in the FAA tests have been the post-crash fuel-fed fires with the fuselage largely intact. Post-crash scenarios have been the focus of FAA work because all accidental fire-related fatalities in the United States in the past 30 years have been due to post-crash fires (Sarkos, 1995; Murray, 1995). Although other scenarios must be considered, a largely intact fuselage (with openings for fire or smoke to enter) has been emphasized by the FAA because (1) the intact fuselage would be more likely to be an impact survivable crash, and (2) direct entry of flames provides the quickest ignition for interior furnishings (Sarkos and Hill, 1989).

The following conditions are considered in the post-crash external fuel-fire scenario:

- *One or more holes in the fuselage; only flame radiation enters.* In this scenario, ignition of interior components is by radiation (no flames), and fire growth is typical of an enclosure fire. The important characteristics include piloted ignition, fire plumes, radiation interactions, and flashover. Fuels for the enclosure fire include interior components and passenger personal items such as clothing and carry-on items.
- *One or more holes (door or rupture) in the fuselage; flames and smoke enter.* In this case, the upper layer of the cabin is quickly vitiated, and thus the thermal decomposition of the exposed panels is different, internal ignition is by direct time contact near fuselage openings. If an escape door (down wind) is opened, the external times may extend along the fuselage ceiling and seriously disrupt escape. If external flames do not extend inside, the fire will develop as an enclosure fire.
- *No holes in fuselage.* The heating of the airplane skin will eventually heat the back side of the interior wall panels, resulting in some degree of aerobic pyrolysis or burnthrough of the skin and insulation systems. Heating of the airplane skin, insulation, and interior is by conduction and high-temperature radiation. The first impediment to escape is toxic pyrolysis products. Once internal ignition occurs by high temperature or a melted hole, the further fire spread is a typical enclosure fire. Effective hull protection that stays in place could provide ample egress time, and a comparison of the time to appreciable heating and degradation of the wall panels with the time for evacuation is important.

In-Flight Fires

Although there have been in-flight fires, in-flight scenarios have caused only a small fraction of fire deaths (Sarkos and Hill, 1989; Sarkos, 1995). For example, during the period from January, 1974, through September, 1989, 892 persistent fire or smoke events were reported for either on-ground or in-flight conditions, with 558 of these occurring in flight (Reynolds et al., 1991) of all of the reported persistent fire and smoke events, 20 progressed to the level of an accident with 9 occurring in flight. Fatalities resulted from 6 of the accidents. In-flight fire statistics are summarized in Figure 3-1. In general, in-flight fires that have resulted in fatalities have started in inaccessible areas of the interior. In response,

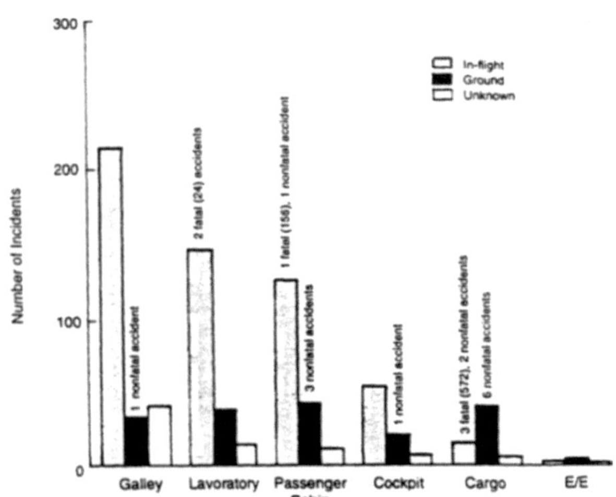

Figure 3-1 Summary of in-flight and on-ground smoke and fire accidents (numbers in parentheses indicate number of deaths in fatal accidents).
Source: Reynolds et al, (1991).

the FAA has pursued research in hidden fire protection, enhanced emergency smoke venting, fire detection and advisory systems, and electrical arc tracking (Sarkos and Hill, 1989).

The following conditions are considered in in-flight fire scenarios:

- *Fire starts within the passenger compartment or lavatory.* This is generally not considered serious since the fire is usually detected and extinguished before it can spread.
- *Fire starts in inaccessible area.* In-flight fires in inaccessible areas can be a serious concern because of the potentially long periods before passengers can be evacuated. For such a fire, the tenability of the cabin and the airworthiness of the aircraft could have to be maintained for up to three hours (for long overseas flights). Inaccessible areas include baggage compartments, concealed spaces behind the cabin liner, avionics compartment, electrical power center, battery compartment, cable trays, cable bundles, air-conditioning components, decompression system, and secondary or attachment structures.

There is a potential for ignition in concealed spaces from an electrical fault or an overheated wire. The nearby materials would be exposed to a sustained but small hot spot or flame. The fire resistance of current materials is adequate to survive this threat. A common cause of in-flight fires is the ignition of materials that are not part of the aircraft design. Examples include trash, grease, oil, dust, and other accumulated matter left during maintenance operations. However, a catastrophic fire cannot generally be produced by a small amount of burning mass unless ignition and flame spread occur on the interior materials near the contaminants. Therefore, ignition, flame spread, and heat release rate of the interior materials are important fire properties. Since the atmosphere in interior areas outside of the passenger cabin is combined with a significant amount of cabin atmosphere and ventilated to the outside of the aircraft, a small amount of pyrolized gaseous products should not reach toxic level in the passenger cabin.

Cargo (or baggage) compartment fires are complex compartment fires. Preventing a fire that may start in a cargo compartment from igniting other areas is very important. In the current jet airplane fleet almost all cargo compartments in passenger airplanes require compartment liners that are airtight and very effective fire barriers. Compartments less than 2,000 cubic feet do not have detectors or fire-suppression agents, and rely on their airtightness and small size to use up the available oxygen in the compartment to suppress a fire before it grows. Compartments larger than 2,000 cubic feet have smoke detectors and use fire-suppression agents and their airtightness to suppress fires. For larger cargo compartments such as those in freighters, one of the procedures used to fight fire is to release the pressure inside the pressure shell to that of the high-altitude atmosphere so that the oxygen content of the compartment is greatly reduced, thus slowing or extinguishing the fire. Fire stops and compartmentalization are important in containing hidden fires.

Fire Dynamics and Fire Load

Fire dynamics is the study of the interrelationships of the basic components in a particular environment in a fire situation. Some of the obvious components affecting fire dynamics are the materials, their characteristics and geometry, and the dynamics of the environment—ambient conditions, environment boundary conditions, occupancy status, ignition sources, fuel loads, and oxidizer quantity and sources.

Important characteristics that influence fire dynamics are ignition temperature and flammability of the component and of the constituent materials. Component geometry can greatly affect ignition and flame spread. Thicker materials take longer to reach their ignition point than thinner materials. The configuration in which the material will be used (e.g., vertical or horizontal, or in a wall or corner) will influence the fire performance. The amount of material exposed is also significant in that a flammable material may be less prone to ignition if is surrounded by nonflammable or less-flammable materials. For example, materials that are located near fuel loads or ignition sources will be more likely to receive direct flame impingement if those fuels are the source of the fire. Other factors that affect the fire resistance of a material are surface roughness and material porosity or susceptibility to wicking.

Generally, fire resistance of a component is improved with:

- increased material thickness (improves resistance to ignition and flame spread),
- decreased exposure of surface area to the environment and increased mass to surface area,
- lower physical location (closer to the floor) in a compartment,
- smoother surface and higher density, and
- higher resistance to ignition and lower flame-spread characteristics.

Based on the considerations described above, it is possible to add up the fire load if the heat release on complete (or partial) combustion of the interior materials is computable (Hill, 1993). The estimation of expected toxic content is less accurate because their production depends on rate of heating, vitiation, and perhaps other conditions which are largely unknown (Levin et al., 1985).

A potentially large, and difficult to define, fire load includes those items that are not considered a permanent part of an aircraft interior, including passenger clothing, carry-on items, and baggage. The passenger materials (papers and clothing) in the seats readily ignite and enhance the ignition of otherwise relatively safe seating. Once the fire has grown to produce a significant ignition source at the ceiling, stowage bin doors that may have been opened on impact or post-crash expose the interior materials which begin to pyrolyze and greatly enhance the ceiling flame extension. Estimates of fire loads are not adequate unless passenger clothing and baggage are included.

FIRE MODELS

An integrated computer modeling capability is necessary to evaluate and compare materials and predict their response in fires. Materials ignition and flame-spread models can be used to predict a material performance. Ignition and flame-spread models along with materials combustion considerations are described in Chapter 4.

Global fire models can be used to improve the understanding of fire scenarios, to evaluate small-scale material test results and provide an estimation of full-scale performance, and to aid the development of fire performance goals. It is necessary to better understand the basic phenomena of fire and smoke propagation within enclosed spaces to provide the needed tools to establish performance goals and evaluate new fire-resistant materials. This requires the development of ignition and flame-spread models for use in a fire-growth model to understand combustion, fluid mechanics, and heat and mass transfer.

Many fire protection computer models are currently available. A recent literature survey identified 74 models from 13 countries (Miller and Friedman, 1992). The two general types of models that are zone models and field models. They each have advantages and disadvantages, and hence are used for different applications. A description of the status of currently available fire models follows. A more detailed review is included in Appendix C.

In *zone models*, compartments are subdivided into control volumes or zones. All quantities of interest are uniform within each zone; conservation of mass, energy, and momentum is applied to each zone using algebraic representations or ordinary differential equations. The advantages of zone models are the low memory requirements, speed, ability to represent large structures, structured output, and ease of use. The need for a priori assumptions and the inaccuracies that may result can also be viewed as a disadvantage. To date, zone models have considered relatively simple, frequently encountered fires that occur in rectangular rooms consisting of flat horizontal fuel beds, vertical walls, perhaps a ceiling, and furniture. Although the required modifications would not be difficult, they have not been equipped for irregular shape enclosures, nonflat ceilings, special fire-geometry interactions, or door-influenced internal air movements.

Field models are based on the numerical solution to a set of partial differential equations representing conservation of mass, energy, and momentum for each chemical species. Field models are usually based on the division of a volume into a large number of computational cells and the application of finite difference techniques to solve for the appropriate quantities in each cell. These models can be used to provide very complex solutions and are often based on first principles, therefore requiring fewer assumptions than zone models. Field models are often used for *ab initio* calculations, can provide a more complete representation of the fire than zone models, and can make new phenomena easier to add. The disadvantages of field models are the requirements for large memory and set-up time, extensive interpretation, and the need for subgrid models (such as turbulence models) if the grid resolution is not sufficiently small. This type of model is computationally intensive and is generally only used when details of the fluid flow are needed.

Toxicity models describe the toxic potency of fire atmospheres based on the toxicological interactions of the main combustion gases present (Hartzell, 1994; Levin et al., 1995; Purser, 1995). Rather than designate a specific combustion system, investigators have the flexibility of designing or choosing a system that will simulate conditions relevant to their fire scenario. One of the models, the N-gas model, is an empirical mathematical relationship containing six gases—carbon monoxide (CO_2), carbon dioxide (CO), low oxygen (O_2), hydrogen cyanide (HCN), hydrogen chloride (HCl), and hydrogen bromide (HBr) (Levin et al., 1987a, b, 1995). Results using the N-gas method have shown the good predictability of this approach (Braun et al., 1988, 1990, 1991; Babrauskas et al., 1990, 1991a, b). Considerations in modeling and testing of toxicity are included in Appendix D.

Fire-hazard assessment models combine zone and field models with submodels for fire endurance, activation of thermal detectors or sprinkler systems, generation of toxic gases, evacuation, and survival models. However, current hazard models were not designed to be applicable to the cylindrical geometry of aircraft interiors or the particular fire scenarios described earlier in this chapter. Thus, their use to model an interior fire on an aircraft will take them outside their domain of applicability.

Today's supercomputers, with their extremely high computational speed and massive storage capability, offer greater opportunity for computer modeling of fires. The arrays of differential equations (either ordinary or partial) that govern the fire phenomena can now be solved numerically. The first models were simple, but current models are building on the older models, incorporating more phenomena and producing more accurate results. As each new submodel (such as a combustion or gas radiation model) is added, the quality of the numerical solutions improves. To apply modeling tools to aircraft fires, modifications need to be made to account for the geometry and fire scenarios involved.

As in all fire models, the definition of the fire will depend on the materials that are burning. New aircraft interior candidate materials will need to be extensively characterized to obtain the data necessary to input into these models.

TESTING AND FIRE-SAFETY ASSESSMENT

Several important properties should be considered when analyzing the burning process. The behavior of a given material is dependent not only on the properties of the fuel, but also on the fire environment in which the material is exposed. For a fire to occur, a fuel source, sufficient oxygen, and sufficient heat are all necessary.

The material properties can be determined through small-scale testing in a laboratory environment. Since small-scale fire tests do not reproduce a fire, most tests do not reflect the hazard a material presents in an actual fire, but provide an indication of the behavior of the materials in an actual fire and a common point for comparison of materials. No single metric, and hence no one test method, is adequate to completely evaluate the fire hazard of a particular material system. For example, the testing procedure for evaluating composite material systems for naval submarine interiors (DOD, 1991) includes oxygen-temperature index, flame spread (ASTM E-162), ignitability (ASTM E-1354), heat release (ASTM E-1354), smoke obscuration (ASTM E-662), combustion gas generation (ASTM E-1354), and toxicity (N-gas method).

Tewarson (1995) and Quintiere (1995) present typical overall characteristics of combustible materials that can presently be used to empirically rank the fire resistance of materials. Advances in the state-of-the-art testing for fire characteristics of materials and the associated developments in the field of mathematical fire modeling make quantitative evaluation of fire hazard feasible. Development of test procedures that allow scaling (e.g., calorimeters and lateral flame-spread methods from which combustion properties can be measured) enable prediction of the behavior of a material under many fire scenario conditions. For simple fire scenarios, hazard variables such as temperature, visibility, toxicity, and corrosiveness of smoke can be related to material properties, such as heat of combustion, heat release rate, smoke particulate yield, smoke extinction coefficient, and yields of combustion gases (e.g., CO_2, CO, low O_2, HBr, NO_2, HCN, and HCl). Determining the properties of these materials can help to establish the principle flammability characteristics—ignitability, flame spread, rate of heat release, and smoke and fire-gas production—discussed earlier in this chapter.

Two additional material parameters that are useful for material fire-hazard comparison are mass-loss rate and oxygen-temperature index. Mass-loss rate does not directly affect fire growth, but the measurement can be used to calculate the effective heat of combustion and, along with smoke obscuration data, is important for fire-growth-model calculations. Oxygen-temperature index is a measure of the percentage of oxygen required for a material to continue to burn at specific temperatures. As the temperature of a material increases, combustion of the material requires less oxygen. An oxygen-temperature profile can be obtained for the material.

Toxicity screening tests for both acute and delayed effects are needed to evaluate the combustion products, including irritant gases of any newly proposed aircraft interior materials and products. Tests should be simple, rapid, inexpensive, use the least amount of sample possible (since, in many cases, only small amounts of the developmental material may be available), use a minimum number of test animals, and have a definitive toxicological endpoint for comparison with other material candidates.

While faulty judgment and incapacitation of passengers in an aircraft fire are significant causes of worry since these conditions can prevent escape and cause death, they are complex endpoints that cannot be directly measured. Death of experimental animals (e.g., rats) is a more definitive and easily determined endpoint and can be used to compare the relative toxicities of alternative materials. Using lethality as the sole endpoint assumes that materials with greater toxicity based on a lethality endpoint will also cause more severe incapacitation and impairment. The number of experimental animals needed for such tests can be significantly reduced by utilizing one of the predictive mathematical models developed for combustion toxicology such as the N-gas model (Levin et al., 1995). Appendix D includes a detailed discussion of toxicity testing and modeling.

Small-scale fire tests have historically been used as a means to screen materials and rank them on a relative basis. Current small-scale tests used for regulatory pass/fail criteria leave much to be desired in terms of being practical, rigorous, well-defined, and repeatable by interlaboratory and intralaboratory equipment or procedures. Progress is being made within the fire research community, including a substantial amount of activity sponsored and encouraged by the FAA (e.g., the International Fire Test Working Group), but additional effort is needed to continue to improve test and analysis methods.

The results of any small-scale flammability test method for a material must be directly relevant to the fire scenario. For example, one possible aircraft fire scenario (described earlier in this chapter) is a fire caused by crash landing in which a large fuel fire occurs outside of an aircraft and penetrates through an opening (or openings) to an aircraft cabin. A critical factor in the escape of passengers is whether (or when) flashover occurs. As described earlier in this chapter, ignition, flame spread, and burning of interior materials might occur. Therefore, the small-scale test method should measure properties of the materials such as piloted ignitiocabin under the expected fire scenarios and to validate small-scale and theoretical models. Detailed measurements of temperature and concentration distribution of chemical species, flow velocity, radiant flux, and records of fire growth using video cameras are needed. The aircraft test configuration should be as realistic as possible, including as many of the interior components described in Chapter 2 as possible (e.g., ducts, wiring, seats, carpets, dividers, windows, and doors). Opening of doors should be considered as one of the parameters. Since the size of airplanes varies significantly, the effect of cabin size on fire-growth rate should be characterized so that optimum test facilities can be developed. Theoretical modeling of these full-scale fires (all types of models, such as zone and field models) should be conducted in close collaboration with the full-scale tests. The comparison of the experimental results with the predicted data would provide not only model validation but also guidelines regarding what experimental measurements are required and where they should be measured.

major toxic gases, burn-out time, flame-spread characteristics, and pyrolysis temperature and provide data to deduce global heat of vaporization properties. Much research has been aimed at developing means to predict the likelihood of flashover from laboratory-scale fire measurements of flammability characteristics and reduced-scale physical modeling (Pitts, 1994).

Rather than being used exclusively as pass/fail screening tests, small-scale tests should be used to measure flammability properties of the materials that can be used as an input to theoretical models to predict fire hazard, described in this chapter and in Appendix C. Since the amount of sample may be limited, especially when testing new experimental materials, the small-scale tests should be designed for as small a sample as possible. However, sample size must be adequate to generate a turbulent flame, maintain a small-edge surface area to the surface area ratio, and allow measurement of flame-spread properties.

Validation with Full-Scale Testing

Aircraft fires are extremely complex for a number of reasons. Aircraft materials are complex mixtures of various polymers, and aircraft components are combinations of a number of materials. The chemical reactions that occur in the solid, liquid, and gaseous state of these materials are imperfectly known, and the gaseous species after pyrolysis and their subsequent reactions have not been fully clarified even in the simplest cases. In a real fire, the turbulent fluid motions, including circulation and mixing with air, further complicates the chemistry and resultant radiation production. Fire phenomena need to be better understood and characterized before computer models can be substantially improved.

Systematic full-scale fire tests are needed to understand important physical processes such as the flow pattern, smoke movement, and fire growth in an aircraft cabin under the expected fire scenarios and to validate small-scale and theoretical models. Detailed measurements of temperature and concentration distribution of chemical species, flow velocity, radiant flux, and records of fire growth using video cameras are needed. The aircraft test configuration should be as realistic as possible, including as many of the interior components described in Chapter 2 as possible (e.g., ducts, wiring, seats, carpets, dividers, windows, and doors). Opening of doors should be considered as one of the parameters. Since the size of airplanes varies significantly, the effect of cabin size on fire-growth rate should be characterized so that optimum test facilities can be developed. Theoretical modeling of these full-scale fires (all types of models, such as zone and field models) should be conducted in close collaboration with the full-scale tests. The comparison of the experimental results with the predicted data would provide not only model validation but also guidelines regarding what experimental measurements are required and where they should be measured.

4

Development of Candidate Materials for Future Interiors

The goal of the current fire-resistant materials research at the Federal Aviation Administration (FAA) is to provide an "order-of-magnitude" improvement in fire resistance compared with current materials. While the goal of an order-of-magnitude improvement is difficult to define considering the multitude of performance metrics and fire scenarios, the ultimate FAA goal to "eliminate fire as a cause of fatalities" (FAA, 1993:1) requires substantial improvements in materials. To achieve such an ambitious goal will require a more fundamental understanding of polymer burning processes and the effects of materials composition, structure, and properties on flammability.

This chapter describes the polymer combustion process and the factors influencing polymer flammability; identifies promising areas for materials research and development; describes factors that will affect selection and application of new materials in a manufacturing environment; and describes modeling techniques that could aid in the understanding of polymer combustion processes and molecular design.

COMBUSTION OF POLYMERS

Polymer combustion is a complex process consisting of a sequence of events including thermal degradation, char formation, transport of degradation products, ignition, and fire growth. An understanding of the factors that influence each event is critical in the development and characterization of new fire-resistant materials.

The amount of energy absorbed by a polymeric material exposed to an external heat source depends on the level and the spectral characteristics of the radiant flux, the in-depth absorption characteristics of the material, and its surface reflectance with respect to the emission spectrum of the incident radiation (Hallman et al., 1978; Kashiwagi, 1981). If the effective absorption coefficient of the material with respect to the thermal radiation is small, a large amount of the material beneath the surface is heated, which slows the rate at which the material approaches its degradation temperature range. However, if the effective absorption coefficient is large, most of the radiation is absorbed close to the surface, and a thin layer of the material is rapidly heated to its degradation temperature range.

Thermal Degradation

When temperatures near the material surface become high, thermal degradation reactions occur and small gaseous degradation products are evolved. The majority of the evolved products are combustible, and their chemical composition depends on the chemical structure of the polymer combined with the degradation conditions. Although the global thermal degradation mechanisms of many different polymers are understood, detailed degradation reaction mechanisms are still not well understood, and degradation mechanisms are especially complex in the case of combinations of materials.

The thermal degradation mechanisms of thermally stable engineering plastics are difficult to understand in detail. Generally, these materials form crosslinks and condensed ring systems that lead to char during degradation. Although only a limited number of analytical methods are available to study changes in chemical structure of the polymer residues because charred or crosslinked samples do not dissolve in solvents, these have not been extensively used and much remains to be done.

Char Formation

Generally, polybutadiene polymers, polyacrylonitrile, poly (vinyl chloride) and many aromatic and heterocyclic backbone polymers can form char under the appropriate conditions. Common to the pyrolysis of all these polymers is the formation of conjugated multiple bonds, transition from a linear to a crosslinked structure, possible formation of ring structures, and an increase of the aromaticity of the polymer residue. General features of the pyrolysis and char of polymers containing aromatic carbon or heterocyclic links in the main chain of the polymer structure have been derived (van Krevelen, 1975; Aseeva and Zaikov, 1985). These features include:

- thermal stability and char yield increase with the relative number of aromatic groups in the main chain per monomer unit of the polymer chain, and
- thermal stability of heterocyclic polymers increases with the aromatic content of the heterocycles.

Since char is mainly composed of carbon and hydrogen, decreased amounts of carbon and hydrogen are released from char-forming materials to the gas phase as combustible gaseous products. Furthermore, since thermal conductivity of char is generally much lower than that of a polymer, a char layer acts as an excellent thermal insulation layer to protect virgin polymer underneath and as barrier to the passage of these gases. However, the detailed chemical reaction steps to form char as well as the chemical structure of char are not well understood. Studies are urgently needed to understand these reactions so as to better enhance the char formation rate and its amount.

Transport of Degradation Products

As a thermal wave penetrates into the interior of a polymer, a highly complex generation and transport of degradation products occurs from the interior of the polymer outward through a strong viscosity gradient. The viscosity gradient has a significant influence on transport behavior. It appears that the transport process of the sub-surface degradation products supplies a majority of degradation products to the sample surface. For char-forming materials, sub-surface degradation products, which are dominant in this case, are transported through the many cracks that form in a hard and porous char. However, if an intumescent char is formed instead of hard, porous char, the transport of the sub-surface degradation products becomes more complex. The effectiveness of the thermal insulation of a char appears to depend on its physical structure. An intumescent, foamy char tends to have better insulation characteristics than a hard, brittle, dense char.

An important macroscopic transport process for thermo-plastic polymers is melting and dripping during burning. Since an aircraft interior consists essentially of a floor, two walls, and a ceiling, if thermoplastic materials on the walls and the ceiling are heated to well above their glass transition temperatures or melt temperatures (for crystalline polymers) by the thermal radiation coming through an opening such as a door, there could be a significant amount of melting and dripping of interior materials. Although such behavior might be helpful in certain fire scenarios, here it appears to increase the hazard; melting and dripping from wall and ceiling panels may interfere with the evacuation of passengers and crew, as well as enhancing fire growth.

Ignition

Ignition occurs after sufficient amounts of combustible degradation products reach the gas phase. Generally, non-charring polymeric materials (e.g., polyolefins) degrade below 400°C which is too low to produce nonpiloted ignition (Kashiwagi, 1981). However, char-forming materials (e.g., phenolics) could reach a surface temperature high enough (due to low thermal conductivity of the char) to allow nonpiloted ignition. However, the limiting requirement to cause piloted ignition is a sufficient supply of gaseous, combustible degradation products. Therefore, piloted ignition tends to occur much earlier than nonpiloted ignition (Kashiwagi, 1981).

Fire Growth

After ignition, the growth of fire is determined by the flame-spread characteristics of materials. The total heat release rate of a fire is determined by the integral of the burning surface area times the local burning rate per unit surface area. Local heat release rates are the result of a complex coupling between condensed and gas-phase phenomena. Flame spread depends on continued generation and transport of degradation products to the flame. The generation rate of combustible degradation products is determined by the heat-and mass-transport processes and also by the chemical degradation reactions.

DEVELOPMENT OF MATERIALS

The relationship between materials chemical structure, composition, and fire performance is presently understood on a general, empirical basis (Weil, 1995; Wilkie, 1995). Previous sections of this report described the burning process as it applies to aircraft interiors and identified important properties and analysis methods to understand combustion and to evaluate materials performance. Based on this understanding of the critical issues and requirements and the structure performance relationships, approaches for significantly improving fire resistance by either affecting the pyrolysis rate or degradation product composition, or by changing the gas-phase reaction rates can be identified. This section outlines potential advances in fire-safe materials, including improvements in organic materials, organic/inorganic materials, inorganic preceramic polymers, innovative material systems, and advanced additive approaches. In addition to the advances described in this report, future prospects for improved fire-resistant materials are described in the proceedings of the conference that the committee hosted (NRC, 1995). Reviews are included that describe research and trends in fire resistant polymers (Wilkie, 1995), additive concepts (Weil, 1995), and inorganic and organometallic polymers (Zeldin, 1995).

Approach

Research to develop a fundamental understanding of chemical structure is most likely to contribute to the development of more-fire-resistant products and reduced flammability characteristics of aircraft interior materials (Sorathia and Beck, 1995). A number of fundamental properties of polymeric materials are important in determining behavior under various fire conditions (Pearce, 1986; Troitzsch, 1990; FAA, 1993; Sorathia and Beck, 1993; Tewarson, 1993; Wilkie, 1993; Lyon, 1994; Nelson, 1995). Among the important physical properties are the morphology (especially the degree of crystallinity), crystalline melting temperature (T_m), glass transition temperature (T_g), thermal conductivity, density, and thermal capacity. Chemical factors include the heat of combustion, the heat of gasification, the minimum surface temperature at which a self-sustaining flame can be established, the products of decomposition, the tendency of the material to form char, and the heat release rate.

As described earlier in this chapter, materials that form a char can produce a barrier at the early stage of burning which can insulate the interior material from further decomposition. To be effective, an adherent thermally insulating layer of char must form on the surface of the polymer to retard the feedback of thermal energy to the underlying material and effectively retard its burning. Char formation offers the opportunity of generalizing fire resistance among a variety of polymeric materials. An important goal should be to develop as much mechanistic information as possible about chemical reactions and physical processes that produce effective char. Also, the effect of char formation on toxicity needs to be characterized.

There are three approaches to developing improved fire-resistant polymers. First, engineering polymers, including thermoplastics such as polycarbonate and polyamides (nylon 6, 6 and nylon 6) and thermosets such as unsaturated or vinyl polyester and phenolic could be improved with additives, coatings, or intumescent agents. This could be accomplished in the near term and may be the lowest cost approach, but it is not clear that the significant improvements sought could be attained.

Second, further understanding of the degradation of current high-temperature specialty polymers could lead to new approaches for enhancing the yield and mechanical integrity of char, which usually results in improved fire resistance. Current specialty polymers include thermoplastics such as polytetrafluoroethylene copolymers, polyphenylene sulfide (e.g., Ryton®), polyarylene ether ketones (e.g., polyetheretherketone and polyetherketoneketone), related systems such as polyarylene ether sulfone (e.g., Radel®), liquid-crystalline polyesters, (e.g., Vectra ®), and the polyether imides (e.g., Ultem®). Thermosets such as cyanate ester, bismaleimide, certain polyimides, and polybenzimidazole are also classified as specialty polymers. The modification of current commercial specialty polymers could provide the best performance in the near term (within 10 years).

Finally, new high-performance thermally stable materials, possibly composed of organic/inorganic systems, copolymers, including novel copolymers containing silicon or phosphorous (Nelson, 1995), polymer blends and alloys, and glasses, ceramics, and laminates could be developed to provide additional improvements in fire resistance. This approach has the potential to provide the greatest improvement in fire performance, but is considered a long-term (10 years) development.

As described in Chapter 2, there is a wide range of material systems used to produce aircraft interior components, including fiber-reinforced polymeric composites, adhesives, decorative layers, thermoplastic molding compounds and formable sheets, textiles, rigid and flexible urethane foams, fibrous mats, acrylic or polycarbonate window transparencies, and elastomers. While this study focuses on the long-term challenge of developing better fire-resistant materials, the ultimate challenge is not just to develop improved polymers, but to also develop a range of products to produce components that are applicable in interiors applications.

It will be important to identify alternate markets for new fire-resistant materials before committing to a development program. The small size of the potential market for materials solely for use in aircraft interior components often does not justify the expense to the suppliers of development and qualification. When asked by the committee to identify drivers for (and barriers against) the development and application of improved fire-resistant materials, industry representatives indicated that "the high cost of qualification and certification of a new material for aircraft applications makes embarking on a material development and implementation program risky for both the materials supplier and the aircraft manufacturer, and without alternative uses for new developments to increase utilization, justifying development of new materials for the limited market will be difficult" (NRC, 1995:238).

Thermosets

The performance, processibility, and light weight of honeycomb sandwich structures produced from polymeric composites that use thermosetting resin matrices have made them the material of choice in most current cabin and cargo liner applications. Prior to the development of the 1980s flammability regulations, the predominant matrix materials were epoxy-based. However, the fire resistance of conventional epoxy networks was not adequate for conformance with interior heat release regulations.

To meet the 1990 heat release requirements (65/65) described in Chapter 2, materials suppliers and manufacturers developed phenolic matrix systems with increased char yield relative to previous phenolic systems and the incumbent epoxy systems that provided flammability and heat release characteristics in compliance with the regulations. Phenolic systems, however, exhibit inferior mechanical properties, some difficulties in processing due to the evolution of volatiles such as formaldehyde, and production concerns centered around the health and safety of workers handling the uncured resin. Alternative cure systems (perhaps epoxy novolacs) could be considered.

Future developments in thermosetting organic-matrix composites will focus on improved methods of production, increased strength and toughness, and improved flammability characteristics (primarily through structural variations that increase char yield). Multifunctional cyanate ester systems based on phenol-formaldehyde backbones are currently available and are finding increased usage, especially in electronic components (due to their low dielectric loss). They cure without generation of volatiles and exhibit exceptional thermal stability and high char yield (Das et al., 1990). Related systems have also been toughened to improve fracture strength (Srinivasan et al., 1994).

Like cyanate ester systems, high-performance thermosetting polymers, including bismaleimides and polyimides, and polybenzimidazoles promise very good fire resistance due to their thermal stability and high char yields. However, high processing temperatures, long cycle times, and high monomer costs detract from their commercial potential and their utility. Also, for thermosetting systems, high-temperature stability is gained at the expense of resin toughness. The mechanical properties and durability reflect the brittle nature of many of these systems. Research continues on approaches to toughening high-temperature thermosets through the addition of a discrete second phase, or cure sites that allow chain extension, or development of an interpenetrating network containing high-temperature thermoplastics.

Thermoplastics

Some high-performance thermoplastics, such as the poly (arylene ether sulfones), polyetherimide, polyphenylene sulfide, and the poly (arylene ether ketones) (e.g., polyetheretherketone, polyetherketoneketone), provide attractive fire resistance due to their high char yields and low heat release rates. These materials have found use on current commercial aircraft, particularly in injection molding and thermoformed sheet product forms.

The use of thermoplastic matrix composites has the potential to provide outstanding fire resistance with greater durability than thermosets, without a weight penalty (Diehl, 1993). There are potential processing advantages, including improved surface finish, shorter process cycles, and the ability to form colored and textured surfaces without additional steps. However, the application of high-performance thermoplastic composites has been limited primarily by their high cost, but also by their high melt viscosity, and high processing temperatures and pressures.

Current high-performance thermoplastics exhibit outstanding fire resistance without additives (Lyon, 1994). Further improvements may be possible through the use of fire-retarding additives or highly fire-resistant thin films or coatings. Innovative additive and coating approaches are described later in this chapter.

Organic/Inorganic Polymers

Organic/inorganic polymers, wherein the inorganic portion contains elements such as silicon, phosphorus, and sulfur, could develop into useful fire-resistant materials. The systems that are important here include those that contain phosphorus, silicon, nitrogen, and perhaps metal containing polymers. The use of silicon has already been demonstrated to some extent. It is known that if one synthesizes polycarbonate siloxane block or segmented copolymers, or polyimide siloxane block or graft copolymers, and exposes them to high temperatures, the resulting materials show significant char that is thought to be a residual organosilicate (Noshay and McGrath, 1977). Another suggested mechanism is that certain of the segmented siloxane copolymers can undergo intumescence, allowing for a volume expansion near the fire source. The relatively stable residue of intumescent char can greatly retard the further entry of the flame to the bulk.

Organophosphorus chemistry is another viable route to nonhalogen containing polymeric materials with high char yield and inherent high flame resistance. There are many examples of the phosphorus-containing systems that have produced desirable effects (Deshpande et al., 1970; Zeldin et al., 1981). Phosphine-oxide containing thermoplastic polymers have been demonstrated to be also outstanding candidates for hydrolytically stable resins with char yield (Smith et al., 1991). This approach has been extended recently to polycarbonates and polyamides (Knauss, 1994; Wan, 1994; Nelson, 1995). Mechanistic studies are needed to determine the role of in-chain inorganic systems such as phosphorus or silicon to allow for full realization of the potential of this approach.

Organic/inorganic hybrids including ketone-phosphorus combinations (Calvert, 1994), as well as copolymers containing phosphorus and sulfur and polyimides containing nitrogen and phosphorus, also show promise. These materials can be synthesized from a variety of different metal alkoxides and organic polymers to create a very fine morphological structure wherein both inorganic oxides and inorganic polymers can coexist. These materials have been referred to as ceramers (Wilkes et al., 1990), ormosils or ormocers (Schmidt, 1988), and polycerams (Boulton et al., 1990). Most of the work to date has been with silicon alkoxides in combination with functionalized short-chain oligomers. Oxides of titanium and zirconium have also been investigated. Further work in this area could lead to important new fire-safe materials. At this point, the high modulus materials have been generated mostly at the expense of developing extreme brittleness (Coltrain et al., 1993). However, there have been isolated reports of siloxane-modified silicates (Spinu and McGrath, 1994) that exhibit more attractive mechanical properties.

Novel blends of phosphate glass with engineering polymer that have similar melt characteristics have been produced (Quinn and Beall, 1992). These materials are intended to have the capability to be processed using standard thermoplastic processes. Work is needed to evaluate the performance and processibility of these materials.

Inorganic Materials

Highly inorganic materials, including inorganic and organometallic polymers offer substantially improved thermal stability and fire resistance compared with more conventional systems. Examples of inorganic and preceramic polymers include polysiloxanes, polysilazanes, polycarbosilanes, polysilanes, and polyphosphazenes (Zeldin, 1995). These polymers have the potential for extremely fire-resistant behavior, however the high cost and limited processibility have restricted their use to specialty areas such as electronic parts. While the potential for long-term benefit is great, it is not clear how soon they could be developed for practical applications in aircraft interiors. Application of inorganic or organometallic polymers may depend on advanced concepts such as thin films or blends as described in the following section.

New and Modified Materials

Since cost is such a critical issue in materials applications on commercial transports, technologies need to be developed to make advanced concepts for enhanced fire performance, such as thin films, blends, coatings, intelligent materials and nanostructures, more practical.

One promising physical approach is the use of a protective coating over a polymer substrate. The coating could be a highly thermally stable polymer layer (e.g., inorganic polymers) or one that generates an intumescent char layer when exposed to external heat. The thickness of the coating and its adhesion to the base material would determine how long it protects the bulk polymer substrate.

A similar concept is to include a heat-sink additive such as aluminum trihydrate using imbedded microcapsules that are micron size (Khalturinskii and Berlin, 1990). Water or flame retardants can be included inside of the capsules. When a polymer having a large number of such capsules is heated to elevated temperatures, the capsules burst and release their ingredients into the flame. The sudden release of water or flame retardants tends to blow off the flame or extinguish it. Since the size of the capsules is so small, their effect on the physical properties of the original polymer is claimed to be negligible.

"Intelligent" materials approaches may provide a means to selectively change the properties of a material to provide improved fire performance. For example, a crosslinking reaction, initiated at a predetermined, elevated temperature could greatly improve thermal stability and char formation. One such material could be polyimides with a phenylethynyl cure site that does not exothermically react until about 350 or 400°C (Meyer et al., 1994).

Another promising new approach is the use of nanostructured materials. Generally, materials with grains or particles 1–100 nanometers across, or with layers or filaments of similar thickness, are considered nanostructured materials. These materials can lead to dramatically improved or altered mechanical, optical, and magnetic properties. The properties of nanostructured materials are determined by a complex interplay among the building blocks and the interfaces between them. While the mechanism of these interactions are not clearly understood, it has been reported that when molecules are intercalated in silicates, their thermal and oxidative stability increases dramatically (Dagani, 1992). The molecules' confinement inside the ceramic lattice apparently protects them from engaging in degradative behavior. It is not clear whether the flammability properties of polymers blended with such nanostructured materials are significantly affected, and feasibility studies are urgently needed. Since this approach does not depend on the polymer base, there might be significant potential.

Flame-Retardant Additives

Flame-retardant (FR) additives have been extensively used to decrease polymer flammability. This is the most common approach since it does not usually affect the in-plant commercial polymer manufacturing procedure. The additives are usually mixed with the manufactured polymer, blended, and then produced as a flammability retarded product (Pearce, 1986).

Various mechanisms for reducing polymer flammability by this route have been reviewed (Pearce, 1986). These mechanisms included volatile-phase active FRs that inhibit the combustion process, condensed-phase active FRs that lead to char or intumescence, and FRs that endothermically lose volatile components (e.g., H_2O). FR additives may function by one or more of these mechanisms in a particular polymer system. To date, most FR additive systems have incorporated either chlorine or bromine, phosphorous, antimony or boron-related compounds, or inorganic hydrates such as alumina trihydrate.

Relatively large amounts of FR are required, and there can be large changes in the balance of properties dependent on whether the FR behaves as a plasticizer or a filler. The search for newer FR systems concentrates on using smaller amounts of FR to allow retention of overall properties of the original polymer; replacement of certain vapor-phase modifying systems such as halogen compounds, since halogen compounds can produce eye and lung irritation during a fire and also cause corrosion problems and post-exposure toxic effects; and decreasing formation of volatile components by increasing condensed-phase reactions leading to enhancement of char and intumescence (Whang and Pearce, 1990).

An ideal FR additive/blend is one that helps to form crosslinks and to enhance cyclization in an originally linear polymer (easier processing) at temperatures well above processing temperature, thus forming char. The mechanisms through which char layers improve fire resistance were described earlier in this chapter. The effectiveness of the thermal insulation of char appears to depend on its physical structure. An intumescent, foamy char consisting of numerous small bubbles tends to have better insulation characteristics than a hard, brittle, dense char. However, char-enhancing FR tend to depend on the polymer structure, and their effectiveness is not as nearly universal as halogenated FR.

Since currently used polymers in the aircraft interior already form a certain amount of char, the enhancement of char amount and the formation of the desirable foamy char in these polymers might not be as difficult as in the case for polyolefin polymers which do not generally form char. However, the level of understanding of thermal degradation mechanisms of engineering plastics is much less than that for polyolefins. This lack of understanding is caused in part by the lack of many analytical tools to characterize the chemical structure of polymer residues that do not normally dissolve in solvents. However, there are techniques such as Fourier-transform infrared spectroscopy and solid-state nuclear magnetic resonance that could be applied to these systems and coupled with microscopy studies.

These approaches and other new FR approaches based on reflectance, endothermicity in the condensed phase, endothermicity in the gas phase, inert gas emission, inhibition of decomposition reactions, char quantity enhancement, rate of char enhancement, char strengthening, catalysis of pre-char chemistry, crosslinking, formation of solid noncarbon barriers, enhancement of the strength of carbonaceous and noncarbon barriers, coating of char, and inhibition of the oxidation of carbonaceous barriers hold the possibility for significantly improved fire resistance (Weil, 1995). Hybrid approaches or synergistic approaches that use multiple additives with differing but cooperating modes of activity in optimized combinations should also be explored (Weil, 1995).

Toxicant Suppressants

A toxicant suppressant is a chemical that, when added to a combustible material, significantly reduces or prevents one or more toxic gases from being generated during thermal decomposition. The resultant gas effluent should be less toxic than that from the untreated material (i.e., the toxic gas whose concentration is being lowered should not be converted to an equally or more toxic product).

Toxicant suppressants are a relatively new concept arising from work demonstrating that the addition of relatively small amounts (0.1–1 percent by weight) of copper compounds to flexible polyurethane (FPU) foam significantly reduced the generation of hydrogen cyanide (HCN), as well as the overall toxicity of the combustion products, when the foam was thermally decomposed (Levin et al., 1988, 1990, 1992). The copper compounds could be added to the foam during formulation without changing the flammability characteristics or physical properties of the foam. Small-scale as well as full-scale room fire tests showed that the toxicant suppressant reduced HCN levels.

The use of melamine-treated FPU is becoming more common because melamine is nonhalogen, inexpensive, and has little effect on processing (Weil and Choudhary, 1995; Weil and Zhu, 1995). It is one of two FPU foams currently allowed in Great Britain. Small-scale tests indicated that a melamine-treated FPU generated at least six times more HCN than an equal amount of a non-melamine-treated foam (Braun et al., 1990, 1991; Levin et al., 1992). The presence of Cu_2O reduced the HCN from the melamine-treated foam by 90 percent. In other experiments, a wool fabric treated with copper generated 50 percent less HCN than the untreated fabric (B.C. Levin, unpublished data, 1995).

Smaller-scale work also showed that the concentrations of HCN generated from the thermal decomposition of thin films of polyurethane at 300 and 400°C decreased when flowed through copper metal films (Jellinek and Takada, 1977; Jellinek et al., 1978). These experiments indicated that the copper is probably acting as an oxidative catalyst which would decompose gaseous HCN into N_2, CO_2, H_2O, and small amounts of nitrogen oxides. Further research is needed to validate this molecular mechanism. The FPU and wool results suggest a more universal effect, namely that treating nitrogen-containing materials with copper compounds could reduce the HCN generated when those materials are exposed to fire conditions. Taking these results one step further, one could develop other toxicant suppressants which when added to materials and products would now prevent or significantly reduce the toxic effluents that are generated when they are exposed to a fire environment.

MANUFACTURING

In addition to the physical and mechanical properties described in Chapter 2 and fire-resistance and toxicity properties described in Chapter 3, improved materials must exhibit processibility and durability necessary for aircraft application and service. Since interior materials do not contribute to the flight performance of the aircraft, component cost, especially life-cycle costs, are critical to their application. New materials must be compatible with manufacturing processes and health and safety considerations (from its initial manufacturing to its final disposal), be available in a variety of product forms, and be cost competitive to be considered for commercial aircraft applications.

Manufacturing Processes

While longer-term developments may have more processing flexibility, new materials developed for aircraft interior applications in the near term need to be compatible with the current methods and equipment. A large investment has been made in manufacturing infrastructure, including automated cutters, presses and autoclaves, and machining equipment. For the most part, the existing methods and equipment offer significant flexibility, allowing a range of materials and constructions to be processed. For near-term applications, developers of new materials must recognize and consider the limitations imposed by existing capabilities.

By far the most prevalent construction found in current aircraft interiors is composite face skins bonded to nonmetallic (usually aramid-reinforced) honeycomb core (Berg, 1995). The extreme light weight of this type of construction will be hard to match with alternate constructions. Hence, thin-skinned honeycomb panels will likely remain the construction of choice for stowage bins and cabin and cargo liner applications.

Manufacturing processes of the future will emphasize cost reduction. Cost reductions in the autoclave or press molding processes currently used in the aircraft industry will likely result from decreasing the high labor intensity of manufacturing steps, including ply preparation (kitting), orientation (layup), molding, surfacing, bonding of decorative surfaces, and part trim. Automation of cutting and ply location processes have shown promise, and improvements could further reduce the cost of these operations. For molding and decorating operations, reduction in the number of process steps and decreasing cycle times hold the most promise. For example, combining molding and application of decorative layers would not only eliminate a bonding cycle, but would preclude the need for highly labor-intensive sanding and surfacing operations currently used to produce a smooth surface. Low-temperature-curing polymers, quick-cure systems, or alternate heat sources (e.g., electromagnetic) could reduce costs by increasing throughput, decreasing energy consumption, simplifying in-process repair, or eliminating the need for part staging to optimize equipment use. Finally, net or near-net shape molding processes could reduce material scrap and eliminate or reduce machining and trim operations.

Unreinforced thermoplastics, including polyetherimide (Ultem), polyetherketoneketone (Declar), and polyarylene ether sulfone (Radel), already find wide use in nonstructural components of current aircraft interiors due to their fire resistance and the advantages of thermoplastic processes, especially surface finish and process cycle time. The most common processes are sheet thermoforming and injection molding.

Developing continuous-fiber reinforced thermoplastic composites to take advantage of thermoplastic processing advantages could provide significant savings in manufacturing costs in future aircraft (Guard and Peterson, 1993; Guard, 1994). Thermoplastic composites could offer the following advantages in future production (Diehi, 1993):

- cheaper tooling, especially for the short production runs typical of commercial aircraft;
- more versatile production methods;
- short production cycles;
- elimination of hand finishing;
- more durable parts without weight penalty;
- integral color, pattern, and texture;
- potential for recycling; and
- better specific fire behavior without loss of durability or appearance.

In general, thermoplastic composites have failed to realize their potential. Most of the manufacturing processing difficulties associated with thermoplastic composites, especially in the production of honeycomb sandwich constructions, result from the high temperatures and pressure required to consolidate these high-viscosity materials. Development of manufacturing processes and compatible materials for co-bonding, secondary bonding, and decorative processes are required before the full potential of thermoplastics can be realized.

Environmental and Health Considerations

The environmental impacts of new fire-resistant or firesafe materials need to be characterized. The historical approach has been to examine the effects of any new process or product on the environment and install mechanisms to mitigate pollution or to clean up the already polluted environment. More recently, environmental issues have been addressed during the design and development phases. There are myriad government and industry programs that emphasize the "design for the environment" approach (OTA, 1992).

There are two aspects to consider in the examination of new chemical compounds and products:

- the effect of the new product on the environment, including adverse effects of the product' s manufacture, byproducts, waste products, or disposal on the environment (air, land, or waters); and
- the effect of the product's manufacture, byproducts, and disposal on the health and well-being of the humans

who are working with, using, or living in the vicinity of the plant making the product or its byproducts or disposal.

Environmental and health and safety issues that influence materials and processes for aircraft interiors include the restrictions on the release of volatile organic compounds, the release of solid and liquid wastes, and the use of toxic or irritant chemicals. Limitations of the release of volatile organic compounds affects materials and processes, including paints and finishes, cleaning and surface preparation, and adhesive bonding. The need to reduce hazardous solid and liquid wastes has significant influence on the disposal of carbon-fiber composite materials,[1] heavy metal compounds such as chromium and cadmium that are used in corrosion prevention finishes,[2] and solutions from manufacturing processes such as cleaning and surface preparation.

The complete evaluation of the toxicity or irritancy of a compound can take as long as five years and is quite costly. The compound may present a concern if:

- the substance may cause carcinogenic effects based ontests or similarity to known carcinogens;
- the substance may cause acutely toxic effects based ontests or similarity to known toxicants;
- the substance may cause serious chronic effects, serious acute effects, or developmentally toxic effects under reasonably anticipated conditions;
- the substance may cause significant adverse environmental effects under reasonably anticipated exposures; or
- concern exists about the health or environmental effects of one or more impurities or byproducts of the substance.

All of the health and safety issues have to be weighted in relation to the potential exposure of individuals (i.e., risk assessment). For example, a new material could be toxic and still be acceptable if exposure can be minimized by taking the appropriate actions during its manufacture and use. The toxicity has to be assessed, however, to determine the extent to which such actions are necessary.

THEORETICAL MODELING OF POLYMER COMBUSTION

Theoretical modeling can be used to predict the influence of material structure on fire performance. A broad range of models of combustion processes have been developed on scales from molecular to intermediate-and full-scale compartment models described in Chapter 3 and Appendix C. The theoretical models discussed in the following sections describe many different aspects of polymer combustion outlined earlier in this chapter. Together, these models could aid in the understanding of polymer combustion processes and in the molecular design of more fire-resistant materials.

Heat Transfer Models

One-dimensional heat transfer models including representation of in-depth absorption in the material are currently available. However, the heat transfer process from the hot surface to the interior of a honeycomb aircraft interior component is very complex (heat conduction occurs through the polymer resin layer, but convective plus radiative heat transfer occur in the honeycomb cells), and an accurate calculation methodology should be developed.

Thermal Degradation Models

The details of the thermal degradation of relatively simple polymers such as polyethylene, polystyrene, and poly (methylmethacrylate) have been studied using molecular dynamic calculations (Nyden and Noid, 1991; Nyden et al., 1992), the Monte Carlo method (Guaita and Chiantoe, 1985; Guaita et al., 1985), and kinetic calculations (Boyd, 1970; Kashiwagi et al., 1989).

Molecular dynamics is a computational technique that has been used to simulate the reaction kinetics of small gas-phase molecules, as well as to study conformational motions in synthetic and natural polymers. More recently, this technique has been applied to investigate the thermal degradation of vinyl polymers and to the design of flame resistant polymers (Nyden and Noid, 1991; Nyden et al., 1992). Although molecular dynamic calculations require a large, high-speed supercomputer, the results simulate the actual dynamic behavior of polymer chains. Molecular dynamics can provide a realistic description of the thermal degradation of polymers and should, therefore, aid in the development of fire-retardant treatments for these materials. Future work will extend these techniques to the combustion of more complex and thermally stable systems such as those used for aircraft interiors.

---

[1] Carbon-fiber composite materials are disposed of as a hazardous solid waste to avoid incineration. Airborne carbon fibers are considered a threat to electrical systems.

[2] Chromium compounds are used in primers in most painted components as well as on metallic support structures, seat tracks, and lavatories. Chromium and cadmium compounds are used in fastener treatments. While the total amount of material may be small, restrictions put on these materials could have a significant influence on future materials selection and development programs.

The Monte Carlo method uses a probability analysis on each set of polymer fractions which consist of a different number of monomeric units. The total set of fractions is equal to the molecular-size distribution of the polymer sample. The change in population in each set is calculated at each time step. Small predetermined macromolecular units, of a predetermined size, that result from the analysis are considered to be evolvable and are counted as lost weight.

Kinetic methods calculate changes in the concentration of polymer molecules and also in the concentration of polymer radicals for each degree of polymerization from the monomer unit to the longest chain. The rate of change in the concentration of each species is expressed by an ordinary differential equation. Numerous differential equations (up to 10,000) are coupled, generally requiring numerical calculations to provide solutions except for specific cases where analytical solutions were derived from approximations. These approximations are based on knowledge of the degradation mechanisms and kinetic expressions for rates of chain initiation reaction, depolymerization, termination reactions, and any other relevant reactions (Boyd, 1970). All these calculations are based on an imposed, spatially uniform temperature and assume no transport processes are involved. There are no theoretical degradation models available that describe the formation of char.

Transport Models

The complex transport processes of degradation products through materials described earlier in this chapter, have been ignored in existing models. Instead, it has been assumed that the degradation products generated in the sub-surface region are instantaneously transported to the sample surface. More progress is needed in modeling complex transport processes.

At present, there are no theoretical models to describe polymer melting and dripping during burning. One needs to understand when melting happens, how fast polymer melt flows, when dripping occurs, how fast dripping occurs, and how much drips. Such modeling is urgently needed to estimate the importance of melt flow and dripping in fire growth.

For the overall modeling of the gasification process, current models generally consist of a time-dependent heat conduction equation with a one-step global degradation reaction. The degradation reaction is approximated at the surface or in the bulk of the polymer. Although mass transport of degradation products from the inside of the sample to the sample surface has been included for wood and bubble growth of coal, these processes are hardly included or even considered for the gasification of polymers except in works of Wichman (1986). Mass transport models for polymers should include nucleation and growth of bubbles and bubble transport processes created by melt viscosity gradient, surface tension gradients, and gravity. Inclusion of such processes, with the addition of char-forming reactions, is also needed for an intumescent char model.

Ignition Models

Generalized ignition processes are reasonably well understood, and two different types of theoretical models are available. Simplified models, the so-called "thermal models," are based on the assumption that the heat-up time of the material is the rate-controlling step and that the chemical reaction time is much shorter. In such a model, an ignition delay time is expressed as the time when the surface temperature of the material reaches its pyrolysis temperature which is taken to be a fixed value. Ignition delay time is correlated inversely with the square of the thermal radiant flux, providing that degradation occurs at a nearly fixed temperature.[3] For a char-forming material, ignition delay time could still be correlated inversely with the nth power of the thermal radiant flux by selecting an appropriate, inferred degradation temperature. Generally, these correlations are derived using experimentally measured results, and the validity of the correlations is limited to the experimental conditions. More-detailed theoretical ignition models solving energy, species equations in the gas phase, and an energy equation in the condensed phase have been developed for the auto-ignition mode (no pilot ignition) in a one-dimensional configuration (Kashiwagi, 1974), at a stagnation point (Amos and Fernandez-Pello, 1988), or for piloted ignition in a boundary layer (Tzeng et al., 1990). However, it is still extremely difficult to quantitatively predict ignition delay time at a specified external radiant flux starting with only information on the chemical structure of a polymeric material. The more complicated models require more-detailed information on thermal and physical properties, especially regarding chemical reactions and their rates. Therefore, it would be unrealistic to attempt to quantitatively predict the ignition characteristics of materials currently used in an aircraft because information on detailed chemical composition of the materials is generally not available. However, it is feasible and desirable to use detailed models as a guideline to improve ignition characteristics during the development of a fire-resistant polymer of known composition.

Fire-Growth Models

After ignition, the growth of fire is determined by the flame-spread characteristics of materials. The most studied case is for well-defined ignition at the bottom of a flat

[3] This condition might be attained for non-char-forming polymeric materials having a high activation energy for a global one-step degradation reaction such as certain poly (methylmethacrylate) polymers.

vertical wall and flame spread upward. A large amount of experimental data and theoretical models are currently available that correlate flame-spread characteristics with materials flammability properties (Hasemi, 1986; Saito et al., 1989; Cleary and Quintiere, 1991). The calculated results show a reasonable agreement with the experimental data (Cleary and Quintiere, 1991). However, the effects of melting or dripping on upward flame spread might be important if thermoplastics are used.

Another important flame-spread configuration for an airplane is flame spread under the ceiling. The limited studies that are available (Agrawal and Atreya, 1992; Atreya and Mekki, 1992) show that unusual flame instability might occur in this configuration. It is extremely difficult to predict the upward flame-spread process if only the chemical structure of the material and its dimensions are provided. Even more-detailed flame-spread models can be developed, including detailed chemical reactions, but their predictive capability would be still semi-quantitative.

Flame-spread and burning characteristics of upholstered furniture have been extensively studied. A theoretical model of fire growth on an upholstered chair, based on detailed radiative shape-factor calculation between the burning area and unburned region with empirical correlation for flame spread, has been reported (Dietenberger, 1992). However, the validity of the model and its accuracy have not yet been well established. These data and models might be of use for aircraft seats to describe how these seats bum in the event of a fire accident.

It is important to be able to predict the occurrence of flashover based on calculated local heat release rates. Given the heat feedback rate from a flame to the material surface, the generation rate of combustible degradation products is determined from the processes of heat-and mass-transport rates processes and also from degradation chemical reaction kinetics. Given the rate of supply of the combustible degradation products and their chemical composition, characteristics of the flame such as flame height and heat release rate are determined. Thus, the direct coupling between the energy feedback rate and the rate of supply of the combustible degradation products determines local heat release rate per unit surface area. Experimental results show that a flame tends to become more optically opaque with increases in flame size, and radiative feedback dominates over convective feedback; also radiative feedback is enhanced for a flame generated by the degradation products having higher aromatic content. At present, it is extremely difficult to calculate radiative transfer in a turbulent flame. There are still uncertainties in how to accurately model turbulence, formation and destruction of soot particulates (the radiation source), and how to calculate radiative heat transfer efficiently.

Since fire in an aircraft cabin is affected by interaction with its surroundings, such as air entrainment through openings or by the interaction of hot ceiling and walls with burning items, compartment fire models described in Chapter 3 and Appendix C are an important element in understanding an aircraft fire.

5

Conclusions and Recommendations Research Opportunities

Aircraft interiors are complex systems that include a number of components—visible items such as flooring, seats, lavatories, ceilings, sidewalls, stowage bins, bag racks, closets, and windows, and items that are not visible to passengers such as ducting, wiring, insulation blankets, and supporting structures. Polymeric materials are predominant, appearing in a wide range of product forms including molded sheet or shapes, composite-faced honeycomb sandwich, textile fibers (fabrics or carpets), foams, sealants, and adhesives. Interiors currently contain materials of varying fire resistance, selected for their particular application and a variety of additional factors such as availability, cost, producibility, and a balance of other useful properties. Environmental safety and health concerns during processing, fabrication, transport, use, and disposal (including ability to recycle), must also be considered. The development of improved fire-resistant materials must consider all these system-related factors in addition to the materials-related flammability characteristics such as reduced heat release, delayed or no ignition, reduced ability to support combustion, and reduced smoke and toxicity.

Materials used in the production of current aircraft interiors, with some exceptions, tend to have better fire resistance than materials used in other transportation systems.[1] Regulatory requirements have been significant driving forces in the optimization of fire resistant polymers and the development of required product forms for aircraft interior applications. Independent programs pursued by industry have also resulted in essentially a new generation of materials that found application in the 747, DC-10, and L-1011, and then a second generation of materials used for the 767, 757, A300-600, and A310. The FAA's heat and smoke release regulations drove improvements to the second-generation materials and to the application of new materials such as more fire-resistant thermoplastics to satisfy specific application needs.

The committee believes that long-term, focused research in fire-resistant polymeric materials can lead to significant improvements in fire performance and safety. To support the development of such materials, advances are required in the understanding and analytical modeling of aircraft fire scenarios, polymer combustion, small-scale characterization tests, and fire-hazard assessments. In addition to the materials' properties, the development process must address the needs of user of the materials. These include the processing and production capabilities of the materials suppliers and aircraft manufacturers; the ability to meet the design, performance, comfort, and aesthetic demands of aircraft interior applications; and compliance with environmental, health, and safety regulations and practices in the manufacture, use, and disposal of aircraft interior components.

The committee focus was on materials technology and enabling design, manufacturing, testing, and modeling capabilities. A comprehensive research program with the goal of improving survival of aircraft accidents would also include aspects of fire- safety and-suppression systems, human factors and behavior in emergencies, and sensor and control development for accident avoidance.

While the committee is confident that significant improvements can be realized in materials performance, the FAA goal of an "order-of-magnitude" improvement in fire resistance is difficult to define because of the multitude of performance metrics and fire scenarios that need to be considered and evaluated. Establishing specific performance goals for materials research based on the current understanding of materials combustion and aircraft fire scenarios is problematic because the data needed to relate materials performance and configurations to observed fire scenarios are not available.

- The committee recommends that materials performance goals for long-term research be established using hazard and risk assessment techniques. These techniques require experimental data from appropriate small-scale tests in conjunction with fire models to predict the expected fire performance and assess the probability of occurrence under realistic conditions, followed by validation tests in the intermediate-and full-scale regime.

[1] Mass transit standards vary and are often difficult to compare with aircraft standards (NRC, 1991). For example, the New York Transit Authority has taken extreme measures to provide minimal fuel to sustain a fire. The October 29, 1995, subway fire in Baku, Azerbaijan, which resulted in over 300 fatalities, provides an example where fire resistance standards were not adequate.

MATERIALS

The committee has identified three key research directions for the development of improved fire-resistant materials:

- Modification of specialty polymers including thermoplastics such as polyetheretherketone, polyetherimide, polyphenylene sulfide, and polysulfone and thermosets such as cyanate esters, bismaleimides, polyimides, and polybenzimidizole. This approach may provide the best performance in the near term (≤10 years).
- Development of new, high-performance, thermally stable materials including organic/inorganic systems, copolymers, polymer blends and alloys, and glasses and ceramics. These materials have the potential for the best performance in the long term (> 10 years).
- Modification of existing engineering polymers including thermoplastics such as polycarbonates, nylons, and polyethyleneterephthalate and thermosets such as phenolics and polyesters. While it is not clear that this approach would lead to the significant improvement in performance sought, this approach may result in significant cost reductions.

A basic scientific understanding of char and intumescence on flammability is crucial to the development of improved materials. Research in char formation should include structural characterization and mechanical behavior (durability) and its relationships to exposure atmosphere, heating rate, chemical derivatization, additives, and coatings. Also the effect of char formation on toxicity needs to be characterized.

The two general technical directions for polymer materials development to improve fire and smoke resistance identified by the committee are incorporation of additives in polymers and synthesis of thermally stable, tim-resistant polymers. Particularly promising approaches are discussed in detail in Chapter 4. These include thin laminated or co-extruded films and blends, coatings and additives (including intumescents), phase-change or temperature sensitive materials, organic/inorganic polymer blends, polymer blends utilizing a high char-forming polymer as an additive, and polymer modifications. Additive approaches include volatile-phase active flame retardants that inhibit the combustion process, condensed-phase active flame retardants that lead to char or intumescence, flame retardants that endothermically lose non-toxic volatile components, heat-sink additives, toxicant suppressants, and combinations of additives that take advantage of synergistic effects (i.e., multiple additives with differing but cooperative modes of activity in optimized combinations).

Recommendations:

- Perform research to improve the fundamental understanding of polymer combustion, including thermal degradation, char formation, intumescence, toxic gas production, and heat effects. Place special emphasis on the characterization of char and intumescence processes.
- Investigate new additive approaches that allow for significant improvements in fire resistance and reduced toxic gas production in current materials.
- Facilitate the development of new or modified polymers with significantly improved resistance to ignition and flame spread. Emphasize the modification of existing specialty polymers to obtain desired properties and the development of new thermally stable polymers or blends.
- Evaluate and prioritize research and technological development efforts to ensure that the new materials will meet end-use requirements. Issues to be considered include costs; the contemporaneous processing and production capabilities of the materials and aircraft industries; ability to meet the design, performance, comfort, and aesthetic demands of aircraft interior applications; and compliance with environmental and health and safety regulations and practices.

COMPONENT DESIGN AND MANUFACTURING

New fire-resistant materials are of little practical value for aircraft interior use if the industrial processing technologies required to manufacture parts are not fully developed and broad-based. Short-and long-term strategies should be developed to characterize new material opportunities for compatibility with existing processes, as well as determining needs for future designs and manufacturing technologies. Short-to mid-range strategies should focus on researching materials that can be produced with existing tooling and manufacturing processes. Long-term strategies should evaluate both materials that can be processed with today's technologies as well as with future technologies. Where improved fire performance can only be achieved with materials requiring new manufacturing processes, materials research and manufacturing process development should be conducted concurrently to ensure smooth implementation.

Research should also be aimed toward developing material constructions with equivalent or lower-weight and simplified processing requirements compared with currently used materials. Acceptance and utilization of new materials would be greatly enhanced where manufacturing and in-use performance advantages are readily achieved and demonstrated. For example, new materials and manufacturing approaches may be able to reduce the number of processing steps and cure time to relieve the current labor-intensive fabrication of honeycomb core sandwich panels. Enhanced thermoplastics could be developed for rapid manufacturing while meeting the other performance requirements described in Chapter 2. Such materials may be easier to rework and recycle and can be more environmentally "benign."

New modular design technologies should be pursued to reduce the required number of parts by integrating components. Minor changes to component designs may also yield improved fire performance, Lightweight films and coatings with improved fire resistance that can be easily integrated into current component constructions such as interior sandwich panels or insulation blankets should be investigated.

Recommendations:

- Prioritize materials research opportunities in terms of compatibility with existing tooling and manufacturing processes. Short-to mid-range programs should focus on materials systems highly compatible with existing manufacturing technologies for a smoother introduction into production. In long-range development where new manufacturing processes are needed, materials research and manufacturing process development should be conducted concurrently to ensure smooth implementation.
- Investigate innovative design and processing concepts such as modular design, fire-resistant films and coatings, new thermoset composite materials and manufacturing approaches to reduce the number of processing steps and cycle times, and expanded use of thermoplastics. These concepts could provide improved fire resistance while reducing manufacturing costs.

FIRE SCENARIOS

A complete understanding of aircraft fires and the responses of materials and components in these fires is required to establish appropriate performance goals and evaluation criteria for new fire-resistant materials. Based on prior experience, two basic fire scenarios have been identified: post-crash fires involving (potentially large) quantities of aviation fuel from ruptured fuel tanks and in-flight fires involving only interior cabin furnishings and passenger-specific items. These scenarios, described in detail in Chapter 3, provide the basis for establishing fire performance behavior and criteria for new materials. However, new aircraft configurations may be significantly different from past designs, and the response of aircraft interiors in these fire scenarios depends on the details of the design. Thus, each aircraft configuration must be analyzed to assess its response. Examples of potential changes that may affect fire scenarios are the variable cross-section, aerodynamically blended fuselage of the proposed High Speed Civil Transport and multiple-deck passenger cabins of very large (800 passenger) sub-sonic transports.[2] Radical changes in aircraft designs, such as a flying wing, would have an even greater effect on fire scenarios but are far less likely to reach production within the 10–20 year time frame considered in this report.

As discussed in Chapter 3, there are several variations of post-crash, external fuel-fire scenarios that need to be considered. Different factors control ignition and fire spread for each variation:

- Case 1: openings in the fuselage and only flame radiation enters. The ignition and flame-spread characteristics of the interior components under the external radiation are critical.
- Case 2: openings in the fuselage allowing flames and fuel degradation products (i.e., hot gases and smoke) to enter. Here the upper layer of the cabin can quickly become vitiated (contaminated with degradation products). Flame spread along the ceiling and toxicity characteristics of combustion products in vitiated atmospheres are critical to survivability in areas away from the openings.
- Case 3: no hull openings. The heating of the airplane skin will eventually overheat the back side of the interior wall panels and insulation, resulting in some degree of aerobic pyrolysis or burnthrough of the skin and insulation systems.

Fuel flammability can overwhelm post-crash fire scenarios. The heat and fire-spread characteristics of a fuel fire can cause materials with outstanding fire resistance to burn readily. Also, as described in Case 2 above, the smoke and hot combustion products of the fuel fire can represent a severe hazard even if interior furnishings do not become involved in the fire. For these reasons, while beyond the scope of the committee's deliberations, the reduction of the fire hazard of fuel is critical in improving survivability in post-crash fires.

In the past, in-flight interior fires have very rarely developed into accident scenarios. Those within the passenger compartment have been detected and extinguished before posing a significant threat and most that began in or around lavatories either were detected and extinguished or self-extinguished. Therefore, in-flight fires in accessible areas within the aircraft interior were not considered in this study. However, in-flight fires in inaccessible areas can be a serious

---

[2] The potential for very large transport aircraft presents challenges in design for fire safety. These challenges include placement and number of emergency exits, effects on detector and extinguishing systems, fire stops and compartment design, and venting.

concern because of the potentially long periods (up to three hours) before passengers can be evacuated. The number of accidents that began as a fire in a cargo compartment is, relatively speaking, extremely small. Nevertheless, recent regulatory upgrades of cargo compartment liners require them to perform as substantial fire barriers to contain possible fires from spreading into the passenger compartment.

The role of the toxicity of combustion products in aircraft accidents needs to be better defined. In post-crash, fuel-fed fire scenarios, smoke originates from at least two sources:

- burning materials inside the cabin (this includes not only standard cabin furnishings, but also combustibles associated with passengers such as carry-on luggage, clothing, and purses), and
- burning jet fuel located in areas where the smoke produced may enter the cabin through hull openings or various doors or escape hatches which must be opened for passenger evacuation.

The FAA determined in its work during the development of the heat release regulation that the toxicity hazard from burning standard cabin furnishings did not become significant until flashover occurred. However, when flashover occurred in the tests, the contribution to hazard from heat alone was shown to become nonsurvivable; hence the contribution of toxicity was essentially not an issue for survivability with the materials tested in such a scenario.

The FAA has not investigated the toxic hazard from the combustion products of jet fuel that can enter the cabin during a post-crash fuel-fed fire scenario. This is an important gap in the current knowledge that must be filled.

Testing and Evaluation

To perform an adequate fire-hazard analysis or flammability assessment of a material, several materials characteristics must be integrated. Experimental data from appropriate small-scale tests should be used in conjunction with fire models to scale the expected fire performance up to realistic behavior, followed by validation tests in intermediate-and full-scale regimes. This type of assessment is sensitive to the end-use application, type of material, and actual fire threats.

Adaptation of current test methods, and, in many cases, new small-scale test methods, are needed to evaluate fire performance characteristics of materials for specified aircraft interior situations and to provide property data for use by modelers to predict component fire performance in expected large-scale fire scenarios. Rather than being used exclusively as pass/fail screening tests, small-scale tests should be used to measure flammability properties of the materials that can be used as input to theoretical models to predict fire hazard. This process requires enhanced interaction between the experimentalist and the modeler to establish that the test procedures are designed to obtain the parameters that the models require. A better understanding of the performance, limitations, and operating principles of existing test equipment and the development of new and better test methods are needed.

Toxicity tests need to be either developed, or currently available procedures need to be appropriately modified, to simulate the fire scenarios of concern in aircraft. Materials present in a fire, including standard cabin furnishings, jet fuel, and passenger items should be decomposed to produce smoke in a manner likely to occur in the fire scenario(s). The gases currently considered in N-gas models—CO, CO_2, HCN, low levels of O_2, HCI, HBr, and NO_2—should be measured and used to predict the LC_{50} values of alternative materials. This prediction can be checked using bioassay experiments to ensure that no toxic gas other than those expected are produced. This approach can be used to compare the toxicity of various experimental materials that could serve in the same end-use. Additional tests procedures need to be developed to examine incapacitation, mutagenicity, carcinogenicity, and teratogenicity from exposures to combustion products.

After small-scale tests, materials and components should be evaluated in tests of increasing scale. An example of such an approach is the military standard for composite materials for submarine applications (DOD, 1991). This standard included a burnthrough fire test to assess the fire resistance of a material and to provide comparison information on its fire containment and propagation (fire, smoke, and fire gases); a quarter-scale fire test to determine the flashover potential of materials in an enclosure when subjected to a fire exposure; a large-scale open environment test to test materials at full size in their intended application under a controlled laboratory fire exposure to determine fire tolerance or ease of extinguishment; and a large-scale pressurizable fire test to test materials in their intended application in a simulated submarine environment under a controlled laboratory fire exposure.

The development of a materials fire-test database can provide a framework to establish performance criteria, evaluate new materials, and predict materials behavior in-service. The steps involved in database development are (1) categorizing and cross-referencing test methods and characteristics and identifying the fire parameters needed for hazard assessment and modeling; (2) compiling existing fire characterization data; (3) obtaining relevant fire-test data through testing programs as they become available; and (4) developing new test methods where needed to provide additional data not available from current methods.

In order to make the database complete, flammability characteristics that must be included are heat of combustion, pyrolysis rate, char depth, heat release rate, char yield, thermal properties, surface temperature at ignition, and heat of gasification. Flame-spread models require good empirical data for these parameters in order to make accurate predictions for time-dependent behavior of materials. Other areas of needed research are material fire performance under unusual conditions such as low oxygen concentrations and low atmospheric pressure or under wind-assisted conditions. Also, current methods do not provide ways to accelerate aging of materials to evaluate retention of low flammability characteristics.

To understand how material systems perform in different configurations and orientations, additional data analysis methodologies are needed (e.g., more heat transfer experimentation, especially for ceiling configuration). Once confidence has been gained on the completeness and accuracy of the materials fire-test database, fire performance of aircraft interior materials can be measured through the judicious implementation of these input parameters in fire models to predict behavior such as rate of flame spread or time to flashover.

Recommendations:

- Develop the science base for small-scale fire performance and toxicity tests, based on expected fire scenarios and verified with full-scale tests, to provide meaningful property data for modeling and materials evaluation.
- Develop a database of materials fire performance properties to provide a means to establish performance criteria, evaluate new materials, and predict materials behavior in aircraft applications correlatable with expected fire scenarios.
- Support technology scale-up through testing on an increasing scale, from small-scale through full-scale testing.

Modeling

An integrated modeling capability for aircraft interior designers could allow the estimation of the performance benefit of various choices of fire-resistant materials and components in aircraft interior applications. Analytical models, ranging in scale from molecular to full scale, are needed to support the development and evaluation of new fire-resistant materials. Models are used to predict materials performance in fires and to assess the fire hazard.

Thermal degradation models that include crosslinking, cyclization, aromatization, and network formation could be used as tools to determine ways to enhance the thermal stability of polymers and to promote char formation during polymer degradation. Intumescent char models based on the formation and growth of bubbles, swelling, polymer melt behavior, and carbonization may be used as tools in optimization of intumescent materials or coatings. The models should be applicable to engineering polymers, specialty polymers, polymers with fire-retardant or toxicant suppressing additives, and polymer blends.

A fire growth model, including submodels to predict ignition, flame spread, heating/pyrolysis/burning, and flame and surface heat transfer, is needed. Current models have greater capabilities to predict smoke spread through multiple compartments than flame spread in both horizontal and vertical directions. Sources that include both thermal or nonflaming and piloted ignition and heating must be considered since both methods produce different modes of preparation of the surface for burning, different products of combustion, different levels of fire intensity, and require different forms of passive fire protection to counteract the source. Once ignited, materials undergo combustion depending on the mode of ignition (energy and exposure time), oxygen availability, and physical and chemical material characteristics.

There are two different types of models: zone models and field models. Zone models are not yet able to include all necessary fuselage characteristics, such as shape (especially ceiling shape), and upper-level fuels, such as combustible ceiling panels and items contained in baggage compartments. Relatively straightforward additions to current zone models should be incorporated. The needed additions are simple in part because, unlike buildings, airplane interiors are now very similar. Current field models solve the Navier-Stokes equations with some type of turbulence model. They need more accurate and computationally efficient radiative heat transfer calculation schemes and turbulent buoyant flow-field calculations, as well as a means to include more realistic condensed-phase and gas-phase chemical reactions.

An area that appears to be inadequate at this time is modeling flame-heat transfer. A good combustion submodel that allows self-consistent burning will provide a global fire model that is less dependent on empirical input for predicting time-dependent heat release rate or mass-loss rate. The advantage of developing a combustion submodel of this type is that the predictive capability of the fire model relies less on the availability of specific material performance data.

For completeness, the mechanisms by which agents suppress combustion or toxic gas production need good model development. Whether acting by physical or chemical means, the mechanisms of how suppression agents work needs to be understood so more effective, environmentally safe fire-suppression agents can be developed. This is especially important since the manufacture and use of most highly effective halo-carbon fire-extinguishing agents is being curtailed because they cause depletion of stratospheric ozone. The same problems associated with modeling ignition have been encountered in suppression modeling. Steady-state conditions do not exist, and changes could be occurring on a micro scale, adding complexity to the modeling process.

Hazard assessment models need to be specially designed to address the aircraft interior, the fire scenarios of interest. Current models do not account for aircraft interior configuration. The combustion of the new materials would need to be tested to generate data for input into the models.

The evacuation of aircraft is very different from that expected in a building or residence and those differences need to be incorporated into hazard models.

As the fire resistance of interior materials improves, toxicity models may need to be expanded to determine the effects on accident survivors of gases likely to be generated if the new fire-safe materials thermally decompose. Other potential effects that should be evaluated are releases of hot water vapor, particulates, and free radicals; the adverse effects of heat on the human body; and the effects of simultaneous exposure to toxic gases and heat. In the long term, the effects of factors such as physical exertion and alcohol may also be incorporated. Instead of test animals (e.g., rats, mice), increased use of in vitro endpoints should also be investigated. Long-term health effects, such as mutagenicity and carcinogenicity, from short-term exposures need to be studied.

Finally, the models under development must be validated to gain the confidence of the design community. This requires a closely coordinated research effort between theoretical model development and intermediate and large-scale validation testing. More emphasis needs to be placed on requiring intermediate and large-scale testing to verify small-scale data and to refine the understanding of the effects of size and configuration of interior components on the fire performance.

Recommendations:

- Develop basic thermal degradation models that are applicable to engineering and specialty polymers and include crosslinking, cyclization, aromatization, and network formation to aid the understanding of polymer stability and char formation. Include both char characteristics and evolved gaseous product properties as key model parameters.
- Develop intumescent char models based on the formation and growth of bubbles, swelling, polymer melt behavior, and carbonization.
- Develop an integrated modeling capability that will allow the estimation of the performance benefit of various choices of fire-resistant materials and components in aircraft interior applications. Work is needed to develop fire-growth, toxicity, and hazard assessment models relating to aircraft fire scenarios.

Long-Term Research Program

The committee believes that the goals of the FAA's research program to develop significant order-of-magnitude improvements in materials fire performance cannot be met with incremental advances or near-term regulatory activity. Rather, substantial advances based on a fundamental understanding of polymer combustion, on accurate aircraft fire scenarios, and on the *systematic* development of materials technology improvements are required. These advancements require a long-term commitment on the part of the FAA working with the aircraft and materials industries and research laboratories.

The uncertainty of new commercial programs, the cost of qualification and certification, and the long time-to-market for new materials tends to discourage suppliers from embarking on materials development efforts. The size of the potential market for materials for use in aircraft interior components often does not justify the expense to the suppliers of development and qualification. Thus, it is important to develop alternate markets for new materials and to apply technology developments from other industries.

Many of the developments that arise from this research will be unique to the issue of commercial aircraft interior fire safety. However, advances in the understanding of polymer combustion, new materials and additive technology, and testing and modeling methods may have applicability to fire safety in other transportation systems such as submarines, ships, mass-transportation systems, automobiles, and buses, as well as commercial and residential buildings. If this long-term research effort is sustained and a coordinated, parallel effort persists in these related areas, significant advances will be made in the understanding of materials fire safety not only for commercial aircraft interiors but also in many other areas where fire safety is a concern.

Recommendations:

- Sustain the effort to develop significantly improved fire-resistant materials as a long-term research program, with clearly stated goals, plans for systematic technology development, and stable financial commitment.
- Continue to follow developments in fire safety in the materials and aerospace industries, as well as in related industries. Coordinate within the U.S. Department of Transportation and with other federal agencies conducting related research, including the National Aeronautics and Space Administration, the departments of Defense, Energy, Transportation, Commerce, and the National Science Foundation.

References

Agrawal, S., and A. Atreya. 1992. Wind-aided flame spread over an unsteady vaporizing solid. Pp. 1685–1693 in Twenty-Fourth Symposium (International) on Combustion. Pittsburgh, Pa.: The Combustion Institute.

Airbus Industries. 1979. Fire and Smoke Toxicity Test Specifications. ATS-1000.01. Toulouse, France: Airbus Industries.

Amos, B., and A.C. Fernandez-Pello. 1988. Model of the ignition and flame development on a vaporizing combustible surface in a stagnation point flow: ignition by vapor fuel radiation absorption . Combustion Science and Technology 62:331–343.

Aseeva, R.M., and G.E. Zaikov. 1985. Flammability of polymeric materials. Advances in Polymer Science 70:172–229.

ASTM (American Society for Testing and Materials). 1983. Test Method for Specific Optical Density of Smoke Generated by Solid Materials. ASTM E-662. Philadelphia, Pa.: ASTM.

ASTM (American Society for Testing and Materials). 1988. Standard Test Method for Surface Flammability of Materials Using a Radiant Heat Energy Source. ASTM E-162. Philadelphia, Pa.: ASTM.

ASTM (American Society for Testing and Materials). 1990. Standard Test Method for Heat and Visible Smoke Release Rates for Materials and Products Using an Oxygen Consumption Calorimeter. ASTM E-1354. Philadelphia, Pa.: ASTM.

ASTM (American Society for Testing and Materials). 1994. Standard Terminology of Fire Standards. Pp. 471–475 in Annual Book of ASTM Standards, Vol. 04.07, ASTM E-176. Philadelphia, Pa.: ASTM.

Atreya, A., and K. Mekki. 1992. Heat transfer during wind-aided flame spread on a ceiling mounted sample. Pp. 1677–1684 in Twenty-Fourth Symposium (International) on Combustion. Pittsburgh, Pa.: The Combustion Institute.

Babrauskas, V., R.H. Harris, Jr., E. Braun, B.C. Levin, M. Paabo, and R.G. Gann. 1990. Large-scale validation of bench-scale fire toxicity tests. Pp. 3–12 in Interflam '90 Fire Safety, C.A. Franks, ed. London: Interscience Communications Ltd.

Babrauskas, V., R.H. Harris, Jr., E. Braun, B.C. Levin, M. Paabo, and R.G. Gann. 1991a. Large-scale validation of bench-scale fire toxicity tests. Journal of Fire Sciences 9:125–148.

Babrauskas, V., R.H. Harris, Jr., E. Braun, B.C. Levin, M. Paabo, and R.G. Gann. 1991b. The Role of Bench-Scale Test Data in Assessing Real-Scale Fire Toxicity. NIST Technical Note 1284. Gaithersburg, Md.: National Institute of Standards and Technology.

Berg, H.D. 1995. Processing and manufacturing of interior components. Pp. 213–219 in Improved Fire- and Smoke-Resistant Materials for Commercial Aircraft Interiors: A Proceedings. National Materials Advisory Board report NMAB-477-2. Washington, D.C.: National Academy Press.

Boeing Commercial Airplane Group. 1977. Aircraft Fireworthiness Interior Design Criteria. Boeing Document D651377. Seattle, Wash.: Boeing.

Boeing Commercial Airplane Group. 1994. Statistical Summary of Commercial Jet Aircraft Accidents: Worldwide Operations 1959–1993. Document D653810-93. Airplane Safety Engineering. Seattle, Wash.: Boeing.

Boulton, J.M., J. Thompson, H.H. Fox, I. Gorodisher, G. Teowee, P.D. Calvert, and D.R. Uhlmann. 1990. Non-linear organic dyes in polyceram hosts. Materials Research Society Symposium Proceedings 180:987.

Boyd, R.H. 1970. The relationship between the kinetics and mechanism of thermal depolymerization. Pp. 47–89 in Thermal Stability of Polymers, R.T. Conley, ed. New York: Marcel Dekker.

Braun, E., B.C. Levin, M. Paabo, J.L. Gurman, H.M. Clark, and M.F. Yoklavich. 1988. Large-Scale Compartment Fire Toxicity Study: Comparison With Small-Scale Toxicity Test Results. NBSIR 88-3764. Gaithersburg, Md.: National Bureau of Standards and Technology.

Braun, E., S. Davis, J.H. Klote, B.C. Levin, and M. Paabo. 1990. Assessment of the Fire Performance of School Bus Interior Components. NISTIR 4347. Gaithersburg, Md.: National Institute of Standards and Technology.

Braun, E., J.H. Klote, S. Davis, B.C. Levin, M. Paabo, and R.G. Gann. 1991. An assessment methodology for the fire performance of school bus interior components. Pp. 855–864 in Fire Safety Science—proceedings of the Third International Symposium, G. Cox and B. Langford, eds. New York: Elsevier.

Calvert, P.D. 1994. Polymers for new materials. Polymer 35(21):4484.

Cleary, T.G., and J.G. Quintiere. 1991. Flammability Characterization of Plastics. NISTIR 4664. Gaithersburg, Md.: National Institute of Standards and Technology.

Coltrain, B.K., C.J. Landry, J.M. O'Reilly, J.M. Chamberlain, G.A. Rakes, J.S. Sedita, L.W. Kelts, M.R. Landry, and V.K. Long. 1993. Role of trialkoxysilane functionalization in the preparation of organic-inorganic composites. Chemistry of Materials 5(10):1445–1455.

Dagani, R. 1992. Nanostructured materials promise to advance range of technologies. C&E News 70:18–24.

Das, S., D.C. Prevorsek, and B.T. DeBona. 1990. Phenolic-triazine resins yield high-performance thermoset composites. Modern Plastics (February) :72–77.

Deshpande, A.P., H.S. Yoon, R. Liepins, and E.M. Pearce. 1970. Some structure-property relationships in polymer flammability: studies on polyethylene terephthalate. Journal of Applied Polymer Science 31:257.

Diehl, R.G. 1993. Applications of continuous fiber reinforced thermoplastics in aircraft interiors. P. 93 in Proceedings of the International Conference for the Promotion of Advanced Fire Resistant Aircraft Materials. Atlantic City, N.J.: Federal Aviation Administration Technical Center.

Dietenberger, M.A. 1992. Upholstered furniture: detailed model. Pp. 479–518 in Heat Release in Fires, V. Babrauskas and S.J. Grayson, eds. New York: Elsevier Applied Science.

DOD (U.S. Department of Defense). 1991. Fire and Toxicity Test Methods and Qualification Procedure for Composite Material Systems Used in Hull, Machinery, and Structural Applications Inside Naval Submarines. MIL-STD-2031(SH). Washington, D.C.: U.S. Department of Defense.

Drysdale, D.D. 1995. Fundamental fire properties of combustible materials. Pp. 37–43 in Improved Fire- and Smoke-Resistant Materials for Commercial Aircraft Interiors: A Proceedings. National Materials Advisory Board report NMAB-477-2. Washington, D.C.: National Academy Press.

FAA (Federal Aviation Administration). 1980. Special Aviation Fire and Explosion Reduction. FAA-ASF-AT-4. Office of Aviation Safety. Atlantic City, N.J.: FAA Technical Center.

FAA (Federal Aviation Administration). 1990. The Materials Fire Testing Handbook. DOT/FAA/CT-89/15. Atlantic City, N.J.: FAA Technical Center.

FAA (Federal Aviation Administration). 1991. Air Safety Research Plan. Atlantic City, N.J.: FAA Technical Center.

FAA (Federal Aviation Administration). 1993. Fire Research Plan. Atlantic City, N.J.: FAA Technical Center.

Guaita, M., and O. Chiantoe. 1985. Molecular mass changes in the thermal degradation of poly-methystyrene. Polymer Degradation and Stability 11:167–179.

Guaita, M., O. Chiantoe, and L. Costa. 1985. Changes in degree of polymerization in the thermal degradation of polystyrene. Polymer Degradation and Stability 12:315–332.

Guard, P.S. 1994. Consideration and selection of aircraft interior materials. Presentation to the Committee on Improved Fire- and Smoke-Resistant Materials for Commercial Aircraft Interiors. Seattle, Wash., April.

Guard, P.S., and J.M. Peterson. 1993. Future material development trends for commercial airplane interiors. P. 313 in the Proceedings of the International Conference for the Promotion of Advanced Fire Resistant Aircraft Materials. Atlantic City, N.J.: Federal Aviation Administration Technical Center.

Hallman, J., C.M. Sliepcevich, and J.R. Welker. 1978. Radiation absorption for polymers: the radiant panel and carbon arcs as radiant heat sources. Journal of Fire Flammability 9:353–366.

Hartzell, G.E. 1994. Smoke toxicity and toxic hazards in fires. ASTM Standardization News 22:38–44.

Hasemi, Y. 1986. Thermal modeling of upward wall flame spread. Pp. 87–96 in Fire Safety Science, Proceedings of the First International Symposium, C.E. Grant, and P.J. Pagni, eds. New York: Hemisphere Publishing Corp.

Hasselbrack, S.A. 1995. Flame-retarding wool textile materials and the evaluation of thermally stable polymers for commercial airplanes. Pp. 165–174 in Improved Fire- and Smoke-Resistant Materials for Commercial Aircraft Interiors: A Proceedings. National Materials Advisory Board report NMAB-477-2. Washington, D.C.: National Academy Press.

Hill, R.G. 1993. The future of aircraft cabin fire safety. Pp. 365in Proceedings of the International Conference for the Promotion of Advanced Fire Resistant Aircraft Materials, R.G. Hill, T. Eklund, and C.P. Sarkos, eds. DOT/FAA/CT-93/3. Atlantic City, N.J.: Federal Aviation Administration Technical Center.

Hottel, H.C. 1959. Review: certain laws governing the diffusive burning of liquids by Blinov and Khudiakov (1957). Fire Research Abstracts and Reviews 1:41.

Jellinek, H.H.G., and K. Takada. 1977. Toxic gas evolution from polymers: evolution of hydrogen cyanide from polyurethanes. Journal of Polymer Science 15:2269–2288.

Jellinek, H.H.G., A. Chaudhuri, and K. Takada. 1978. Inhibition of hydrocyanic acid evolution from polyurethanes during oxidative degradation. Polymer Journal 10:253–268.

Kashiwagi, T. 1974. Radiative ignition model of a solid fuel. Combustion Science and Technology 8:225–236.

Kashiwagi, T. 1981. Radiative ignition mechanism of solid fuel. Fire Safety Journal 3:185–200.

Kashiwagi, T., A. Inaba, and A. Hamins. 1989. Behavior of primary radicals during thermal degradation of poly (methylmethacrylate). Polymer Degradation and Stability 26:161–184.

Khalturinskii, N.A., and A.A. Berlin. 1990. On reduction of combustibility of polymeric materials. International Journal of Polymeric Materials 14:109–125.

Knauss, D.M. 1994. Synthesis and Characterization of New Polycarbonate Material (Bisphenol A Polycarbonate), Hydrolytic Stability, Fire Resistance. Ph.D. Thesis, Virginia Polytechnic Institute and State University, Blacksburg.

Levin, B.C., A.J. Fowell, M.M. Birky, M. Paabo, A. Stolte, and D. Malek. 1982. Further Development of a Test Method for the Assessment of the Acute Inhalation Toxicity of Combustion Products. NBSIR 82–2532. Gaithersburg, Md.: National Institute of Standards and Health.

Levin, B.C., M. Paabo, M.L. Fultz, and C.S. Bailey. 1985. Generation of hydrogen cyanide from flexible polyurethane foam decomposed under different combustion conditions. Fire and Materials 9:125–133.

Levin, B.C., M. Paabo, J.L. Gurman, and S.E. Harris. 1987a. Effects of exposure to single or multiple combinations of the predominant toxic gases and low oxygen atmospheres produced in fires. Fundamental and Applied Toxicology 9:236–250.

Levin, B.C., M. Paabo, J.L. Gurman, S.E. Harris, and E. Braun. 1987b. Toxicology interactions between carbon monoxide and carbon dioxide. Toxicology 47:135–164.

Levin, B.C., M. Paabo, J.L. Gurman, H.M. Clark, and M.F. Yoklavich. 1988. Further studies of the toxicological effects of different time exposures to the individual and combined fire gases: CO, HCN, CO2, and reduced oxygen. Pp. 249–252 in Polyurethane '88, Proceedings of the 31 st Society of Plastics Meeting. Lancaster, Pa.: Technomic Publishing Co.

Levin, B.C., M. Paabo, R.H. Harris, N. Eller, and L. Highbarger. 1989. Copper Additives Decrease the Acute Inhalation Toxicity of the Combustion Products of Flexible Polyurethane by Inhibiting Hydrogen Cyanide Generation, Annual Report, Year 2. Gaithersburg, Md.: National Institute of Standards and Technology.

Levin, B.C., E. Braun, J.R. Shields, and D. Lowe. 1990. Reduction of hydrogen cyanide concentrations and acute inhalation toxicity from flexible polyurethane foam combustion products by the addition of copper compounds: Part III. The effect of copper additives on the flammability characteristics of flexible polyurethane foam. NISTIR 4441. Gaithersburg, Md.: National Institute of Standards and Technology.

Levin, B.C., E. Braun, M. Paabo, R.H. Harris, and M. Navarro. 1992. Reduction of hydrogen cyanide concentrations and acute inhalation toxicity from flexible polyurethane foam combustion products by the addition of copper compounds. Part IV. Effects of combustion conditions and scaling on the generation of hydrogen cyanide and toxicity from flexible polyurethane foam with and without copper compounds. NISTIR 4948. Gaithersburg, Md.: National Institute of Standards and Technology.

Levin, B.C., E. Braun, M. Navarro, and M. Paabo. 1995. Further development of the N-gas model: an approach for predicting the toxic potency of complex combustion mixtures. Pp. 293–311 in Fire and Polymers II, Materials and Tests for Hazard Prevention, G. Nelson, ed. ACS Symposium Series 599. Washington, D.C.: American Chemical Society.

Lyon, R.E. 1994. Advanced Fire-Safe Materials Research Program. DOT/FAA/CT-94/60. Atlantic City, N.J.: Federal Aviation Administration Technical Center.

Lyon, R.E., and T. Eklund. 1993. Federal Aviation Administration advanced aircraft materials research and development plan. Pp. 345–364 in Proceedings of the International Conference for the Promotion of Advanced Fire Resistant Aircraft Materials, R.G. Hill, T. Eklund, and C.P. Sarkos, eds. DOT/FAA/CT-93/3. Atlantic City, N.J.: Federal Aviation Administration Technical Center.

Meyer, G.W., B. Tan, and J.E. McGrath. 1994. Solvent-resistant polyetherimide network systems via phenylethylphthalic anhydride endcapping . High Performance Polymers 6:423–435.

Miller, C., and R. Friedman. 1992. Overview of models: an international survey of computer models for fire and smoke. Journal of Fire Protection Engineering 4 (3): 1–16.

Murray, T.M. 1995. (The Boeing Company) Airplane accidents and fires. Pp. 7–23 in Improved Fire- and Smoke-Resistant Materials for Commercial Aircraft Interiors: A Proceedings. National Materials Advisory Board report NMAB-477-2. Washington, D.C.: National Academy Press.

Nelson, G.L., ed. 1995. Fire and Polymers II: Materials and Tests for Hazards Prevention. ACS Symposium Series 599. Washington, D.C.: American Chemical Society.

NFPA (National Fire Protection Association). 1988. The Fire Protection Handbook. Quincy, Mass.: NFPA.

Noshay, A., and J.E. McGrath. 1977. Block Copolymers: Overview and Critical Survey. New York: Academic Press.

NRC (National Research Council). 1991. Fires in Mass Transit Vehicles: Guidelines for the Evaluation of Toxic Hazards. National Materials Advisory Board report NMAB-462. Washington, D.C.: National Academy Press.

NRC (National Research Council). 1995. Improved Fire- and Smoke-Resistant Materials for Commercial Aircraft Interiors: A Proceedings . National Materials Advisory Board report NMAB-477-2. Washington, D.C.: National Academy Press.

Nyden, M.R., and D.W. Noid. 1991. Molecular dynamics of initial events in the thermal degradation of polymers. Journal of Physical Chemistry 95:940–945.

Nyden, M.R., G.P. Forney, and J.E. Brown. 1992. Molecular modeling of polymer flammability: application to the design of flame-resistant polyethylene. Macromolecules 25:1658–1666.

OTA (Office of Technology Assessment). 1992. Green Products by Design: Choices for a Cleaner Environment. OTAE-541. Washington, D.C.: U.S. Government Printing Office.

Pearce, E.M. 1986. Flame retardants for polymer systems. Pure and Applied Chemistry 58:925–930.

Pitts, W.M. 1994. An Engineering Algorithm for the Estimation of Carbon Monoxide Generation in Enclosure Fires. Paper presented at the Annual Conference on Fire Research, National Institute of Standards and Technology, Gaithersburg, Md. October 17–20.

Purser, D. 1995. Smoke toxicity. Pp. 175–195 in Improved Fire- and Smoke-Resistant Materials for Commercial Aircraft Interiors: A Proceedings. National Materials Advisory Board report NMAB-477-2. Washington, D.C.: National Academy Press.

Quinn, C.J., and G.H. Beall. 1992. Low temperature glass polymer blends. In Chemical Processes of Advanced Materials, L.L. Hanch and J.K. West, eds. New York: Wiley.

Quintiere, J.G. 1995. Fire tests and hazard evaluation. Pp. 45–59 in Improved Fire- and Smoke-Resistant Materials for Commercial Aircraft Interiors: A Proceedings. National Materials Advisory Board report NMAB-477-2. Washington, D.C.: National Academy Press.

Reynolds, T.L., G.E. Grimstad, and C.D. Anderson. 1991. Aircraft Command in Emergency Situations (ACES) Phase 1: Concept Development. DOT/FAA/CT-90/21. Atlantic City, N.J.: Federal Aviation Administration Technical Center.

Saito, K., F.A. Williams, I.S. Wichman, and J.Q. Quintiere. 1989. Upward turbulent flame spread on wood under external radiation. Journal of Heat Transfer 111:438–445.

Sarkos, C.P. 1989. Development of Improved Fire Safety Standards Adopted by the Federal Aviation Administration, Figure 6, Effective Composite Panel Design on Fractional Effective Dose; North Atlantic Treaty Organization Advisory Group for Aerospace Research and Development (AGARD), Conference Proceeding No. 467, Aircraft Fire Safety presented at AGARD meeting, Sintra, Portugal, May 22–26, 1989.

Sarkos, C.P. 1995. Heat exposure and burning behavior of cabin materials during an aircraft post-crash fuel fire. Pp. 25–36 in Improved Fire- and Smoke-Resistant Materials for Commercial Aircraft Interiors: A Proceedings. National Materials Advisory Board report NMAB-477-2. Washington, D.C.: National Academy Press.

Sarkos, C.P., and R.G. Hill. 1989. Characteristics of aircraft fires measured in full-scale tests. Pp. 11–1 to 11–17 in AGARD Conference Proceedings No. 467 on Aircraft Fire Safety, Sintra, Portugal, May 22–26, 1989. Neuilly-Sur-Seine, France. North Atlantic Treaty Organization, Advisory Group for Aerospace Research and Development.

Schmidt, H. 1988. Inorganic and Organometallic Polymers, M. Zeldin, K.J. Wynne, and H.R. Allcock, eds. ACS Symposium Series 360 333. Washington, D.C.: American Chemical Society.

Smith, C.D., H. Grubbs, H.F. Webster, A. Gungor, J.P. Wightman, and J.E. McGrath. 1991. Unique characteristics derived from poly (arylene ether phosphine oxide)s. High Performance Polymers 3(4):211–229.

Sorathia, U., and C. Beck. 1993. Fire Screening Results of Polymers and Composites. Paper presented in the Science and Technology of Fire Resistant Materials, the 39th Sagamore Conference, Sagamore Army Materials Research Laboratory, Watertown, Mass., September 14–17, 1992.

Sorathia, U., and C. Beck. 1995. Fire Screening Results of Polymers and Composites. Pp. 93–114 in Improved Fire and Smoke-Resistant Materials for Commercial Aircraft Interiors: A Proceedings. National Materials Advisory Board report NMAB-477-2. Washington, D.C.: National Academy Press.

Spinu, M., and J.E. McGrath. 1994. Synthesis of modified SiO2 networks: polydimethyl-siloxane-SiO_2 organic-inorganic hybrids via sol-gel chemistry. Journal of Inorganic and Organometallic Chemistry 2(1): 103–128.

Srinivasan, S.A., S.S. Joardar, D.B. Priddy, Jr., T.C. Ward, and J.E. McGrath. 1994. Reactive poly (arylene ether) toughening of cyanate ester networks via thermal and microwave curing. Materials Research Society Symposium Proceedings 347:705–710.

Tewarson, A. 1993. Flammability parameters of materials: ignition, combustion, and flame spread. P. 263 in Proceedings of the International Conference for the Promotion of Advanced Fire Resistant Aircraft Materials. Atlantic City, N.J.: Federal Aviation Administration Technical Center.

Tewarson, A. 1995. Fire properties of materials. Pp. 61–91 in Improved Fire- and Smoke-Resistant Materials for Commercial Aircraft Interiors: A Proceedings. National Materials Advisory Board report NMAB-477-2. Washington, D.C.: National Academy Press.

Troitzsch, J. 1990. International Plastics Flammability Handbook, 2nd ed. New York: Carl Hanser Verlag.

Tzeng, L.S., A. Atreya, and I.S. Wichman. 1990. A one-dimensional model of piloted ignition. Combustion and Flame 80:94–107.

van Krevelen, D.W. 1975. Some basic aspects of flame resistance of polymeric materials . Polymer 16:615–620.

Wan, I. 1994. Synthesis and Characterization of High Performance Polymeric Poly (Arylene Ether)s, Polyamides, Polyesters and Liquid Crystal Polyarylates. Ph.D. Thesis, Virginia Polytechnic Institute and State University, Blacksburg.

Weil, E.D. 1995. Meeting FR goals using polymer additive systems. Pp. 129–150 in Improved Fire- and Smoke

Resistant Materials for Commercial Aircraft Interiors: A Proceedings. National Materials Advisory Board report NMAB-477-2. Washington, D.C.: National Academy Press.

Weil, E., and V. Choudhary. 1995. Flame-retarding plastics and elastomers with melamine. Journal of Fire Sciences 13:104–126.

Weil, E.D., and W. Zhu. 1995. Some practical and theoretical aspects of melamine as a flame retardant. Pp. 199–216 in Fire and Polymers II, Materials and Tests for Hazard Prevention, G. Nelson, ed. ACS Symposium Series 599. Washington, D.C.: American Chemical Society.

Whang, W.T., and E.M. Pearce. 1990. Polymers with improved flammability characteristics. Pp. 226–273 in Fire and Polymers, ACS Symposium Series 425, G.L. Nelson, ed. Washington, D.C.: American Chemical Society.

Wichman, I.S. 1986. A model describing the steady-state gasification of bubble-forming thermoplastics in response to an incident heat flux . Combustion and Flame 63:217–229.

Wilkes, G.L., H.H. Huang, and R.H. Glaser. 1990. Silicon-Based Polymer Science. J.M. Ziegler and F.W.G. Fearon, eds. Advanced Chemical Series 224, 207. Washington, D.C.: American Chemical Society.

Wilkie, C.A. 1993. Fire Properties of Future Materials Candidates. Paper presented in the Science and Technology of Fire Resistant Materials, the 39th Sagamore Conference, Sagamore Army Materials Research Laboratory, Watertown, Mass., September 14–17, 1992.

Wilkie, C.A. 1995. Fire properties of future material candidates. Pp. 115–128 in Improved Fire- and Smoke-Resistant Materials for Commercial Aircraft Interiors: A Proceedings. National Materials Advisory Board report NMAB-477-2. Washington, D.C.: National Academy Press.

Zeldin, M. 1995. An overview of inorganic and organometallic polymers. Pp. 151–164 in Improved Fire- and Smoke-Resistant Materials for Commercial Aircraft Interiors: A Proceedings. National Materials Advisory Board report NMAB-477-2. Washington, D.C.: National Academy Press.

Zeldin, M., J. Witt, and E.M. Pearce. 1981. Synthesis and thermal properties of crosslinkable phosphazene copolymers. Journal of Polymer Science, Chem. Ed., 19:917.

APPENDIX A

Glossary of Terms

Aircraft accident: an occurrence associated with the operation of an aircraft that takes place between the time any person boards the aircraft with the intention of flight until such time as all such persons have disembarked, in which any person suffers death or serious injury as a result of being in or upon the aircraft or by direct contact with the aircraft or anything attached thereto, or the aircraft receives substantial damage (Murray, 1995).

Aircraft interior: the components and systems that are contained inside the pressure shell, that is, the pressurized part of the aircraft fuselage.

Assist (or Abuse) loads: design loads associated with hard use including bumping, pushing, and pulling handles.

Burn length: the distance from the original specimen edge to the farthest evidence of damage to the test specimen due to that area's combustion (FAA, 1990).

Char: carbonaceous material formed by pyrolysis or incomplete combustion (ASTM, 1994).

Controlled flight into terrain (CFIT): aircraft accident scenario wherein the aircraft crashed into the ground while under control of the flight crew.

Crocking: damage of textiles caused by abrasion.

Extinguishing time: the time that the material continues to flame after the flame source is removed (FAA, 1990).

Field models: fire models based on the division of a volume into a large number of computational cells and the application of finite difference techniques to provide numerical solutions to partial differential equations representing conservation of mass, energy, and momentum for each chemical species.

Fire gases: airborne products emitted by a material undergoing pyrolysis or combustion that exist in the gas phase at the relevant temperature (ASTM, 1994).

Fire hazard: the potential for harm associated with fire (ASTM, 1994).

Fire-hazard assessment models: fire models that combine zone and field models with submodels for fire endurance, activation of thermal detectors or sprinkler systems, generation of toxic gases, evacuation, and survival models.

Fire risk: an estimation of expected fire loss that weighs fire hazards associated with various fire scenarios with the probability of occurrence of those scenarios.

Flame spread: a measure of the rate of burning that is derived from the rate of progress of the flame front.

Flashover: the point at which most of the combustible materials in an enclosed space reach their ignition temperatures at essentially the same time, so that the materials seem to burst into flame simultaneously.

Heat release: a measure of the amount of energy evolved by a material when burned (FAA, 1990).

Heat release rate: the rate at which heat energy is evolved by a material when burned. The maximum heat release rate occurs when the material is burning most intensely (FAA, 1990).

Ignitability: ease of ignition (i.e., initiation of combustion; ASTM, 1994).

Intumescence: swelling or foaming of a material upon exposure to a critical level of thermal energy.

LC_{50}: the concentration of a toxic substance that causes death in 50 percent of test organisms in a specified time.

Limit loads: typical flight loads.

Nonpiloted ignition: the initiation of combustion without contact with an external, high-energy source. The surface temperature of a material must become high enough to act as an induced pilot to initiate gas-phase oxidation reactions and attain ignition.

Oxygen-temperature index: a measure of the percentage of oxygen required for a material to continue to burn at specific temperatures.

Piloted ignition: the initiation of combustion as a result of contact of a material or its vapors with an external, high-energy source (ASTM, 1994).

Pyrolysis: irreversible chemical decomposition caused by heat, usually without oxidation (ASTM, 1994).

Smoke: airborne solid and liquid particulates and gases evolved when a material undergoes pyrolysis or combustion (ASTM, 1994).

Specific optical density: a dimensionless measure of the amount of smoke produced per unit area based on light transmittance measurements (FAA, 1990).

Survivable accident: an accident in which the fuselage remains relatively intact, the crash forces do not exceed the levels of human tolerance, there are adequate occupant restraints, and there are sufficient escape provisions (Murray, 1995).

Tapestry: woven decorative fabrics that are bonded to cabin liner panels.

Thermoplastic: polymers that soften and flow upon application of heat.

Thermoset: polymers that, when heated, react irreversibly so that subsequent applications of heat do not cause them to soften and flow.

Toxicity models: describe the toxic potency of fire atmospheres based on the toxicological interactions of the main combustion gases present.

Toxic potency: a quantitative expression relating concentration (of smoke or combustion gases) and exposure time to a particular degree of adverse physiological response (e.g., the death on exposure of humans or animals) (ASTM, 1994).

Ultimate loads: design loads corresponding to limit loads (typical flight loads) multiplied by a margin of safety (typically 1.5).

Zone models: fire models in which compartments are subdivided into control volumes or zones. Conservation of mass, energy, and momentum is applied to each zone using algebraic representations or ordinary differential equations.

REFERENCES

ASTM (American Society for Testing and Materials). 1994. Standard Terminology of Fire Standards. Pp. 471–475 in Annual Book of ASTM Standards, Vol. 04.07, ASTME-176. Philadelphia, Pa.: ASTM.

FAA (Federal Aviation Administration). 1990. The Materials Fire Testing Handbook. DOT/FAA/CT-89/15. Atlantic City, N.J.: FAA Technical Center.

Murray, T.M. 1995. (The Boeing Company) Airplane accidents and fires. Pp. 7–23 in Improved Fire- and Smoke-Resistant Materials for Commercial Aircraft Interiors: A Proceedings. National Materials Advisory Board report NMAB-477-2. Washington, D.C.: National Academy Press.

Appendix B

FAA Research and Development

ORIGINAL MANDATES

Prior to 1958, the responsibilities for aviation safety rule making and air traffic control were divided among the Civil Aeronautics Board, the Civil Aeronautics Administration, and other federal agencies. The Federal Aviation Act of 1958, whose passage was stimulated by a series of midair collisions, established the Federal Aviation Agency within the U.S. Department of Commerce. The agency inherited the existing rule making and air traffic control responsibilities, plus the additional responsibility for controlling military as well as civil traffic. Subsequently, in 1965 the Federal Aviation Agency was renamed the Federal Aviation Administration (FAA) and relocated from Commerce to the U.S. Department of Transportation, which had been recently created to promote nationally integrated and balanced transportation systems.

Since it was originally established in 1958, part of the FAA's responsibilities has involved research and development. Section 312 of the Federal Aviation Act of 1958 originally charged the agency to "undertake or supervise such developmental work and service testing as tends to the creation of improved aircraft, aircraft engines, propellers, and appliances," which essentially limited the FAA to applied research in the testing and evaluation of then-available technology.

NEW MANDATES

The Aviation Safety Research Act of 1988

The Aviation Safety Research Act of 1988 (Public Law 100–591) amended the Federal Aviation Act of 1958 in two fundamental ways. First, i t added certain topics to those on which the FAA was already mandated to conduct research (research and development is addressed in Sections 312 and 316 of the Federal Aviation Act [49 U.S.C. App. 1353 et seq.], namely:

- aviation mainte nance (addresses the aging fleet)

  The (FAA) Administrator shall undertake or supervise research to develop technologies and to conduct data analysis for predicting the effects of aircraft design, maintenance, testing, wear, and fatigue on the life of aircraft and on air safety, to develop methods of analyzing and improving aircraft maintenance technology and practices (including nondestructive evaluation of aircraft structures).
- fire safety (addresses fire containment and fire resistance of engine fuel and cabin materials)

  The Administrator shall undertake or supervise research to assess the fire and smoke resistance of aircraft materials, to develop and improve fire and smoke containment systems for in-flight aircraft fires, and to develop advanced aircraft fuels with low flammability and technologies for containment of aircraft fuels for the purpose of minimizing post crash fire hazards.
- human factors (addresses performance of flight crew, aircraft mechanics, and air traffic controllers)

  The Administrator shall undertake or supervise research to develop a better understanding of the relationship between human factors and aviation accidents and between human factors and air safety, to enhance air traffic controller and mechanic and flight crew performance, to develop a human factor analysis of the hazards associated with new technologies to be used by air traffic controllers, mechanics, and flight crews, and to identify innovative and effective corrective measures for human errors which adversely affect air safety.
- dynamic simulation modeling of the air traffic control system (addresses air traffic capacity and control)

  The Administrator shall undertake or supervise a research program to develop dynamic simulation models of the air traffic control system which will provide analytical technology for predicting air traffic control safety and capacity problems, for evaluating planned research projects, and for testing proposed revisions in operations programs.

The second fundamental change was to establish the Civil Aeromedical Institute, which had formerly been created and operated solely at the discretion of the FAA, as a legislatively mandated arm of the FAA, and to specifically charge it with pursuing research in human factors as a crucial functional responsibility. The Committee on Science, Space, and Technology of the House of Representatives, which was responsible for creating this legislation, cited human factors as a vitally important research area and indicated its concurrence with expert testimony that "if significant improvements are to be made in the technology of transportation safety, they must come at the hands of human factors researchers, or not at all."

In addition, the Act required two further items:

- a national aviation research plan and annual reports to Congress

  The Administrator shall prepare and transmit to Congress, a national aviation research plan. Not later than the date of the submission of the President's budget ... for each fiscal year ..., the Administrator shall review and revise the plan and pu blish and transmit the revised plan to the Committee on Science, Space, and Technology of the House of Representatives and the Committee on Commerce, Science, and Transportation of the Senate. The plan shall describe, for a 15-year period, the research, engineering, and development considered by the Administrator necessary to ensure the continued capacity, safety, and efficiency of aviation in the United States, considering emerging technologies, forecasted needs of civil aeronautics, and provide the highest degree of safety in air travel.

  The Administrator shall report to the Committee on Science, Space, and Technology of the House of Representatives and the Committee on Commerce, Science, and Transportation of the Senate on the accomplishments of the research completed during the preceding fiscal year.

- an FAA research advisory committee

  The advisory committee shall provide advice and recommendations to the Administrator regarding needs, objectives, plans, approaches, content, and accomplishments with respect to the aviation research program.... The committee shall also assist in assuring that such research is coordinated with similar research being conducted outside of the Federal Aviation Administration. The advisory committee shall be composed of not more than 20 members appointed by the Administrator from among persons who are not employees of the Federal Aviation Administration and who are especially qualified to serve on the committee by virtue of their education, training, or experience. The Administrator in appointing the members of the committee shall ensure that universities, corporations, associations, consumers, and other government agencies are represented.

  The chairman of the advisory committee shall be designated by the Administrator.

The Aircraft Catastrophic Failure Prevention Research Act of 1990

Subsequent to the Aviation Safety Research Act of 1988, the Aircraft Catastrophic Failure Prevention Research Act of 1990, which was incorporated in the Omnibus Reconciliation Act of 1990 (Public Law 101–508), further amended Section 312, charging the FAA "to develop technologies and methods to assess the risk of and prevent defects, failures and malfunctions of products, parts, processes, and articles manufactured for use in aircraft, aircraft engines, propellers and appliances which could result in catastrophic failure of an aircraft."

FAA RESPONSE TO THE NEW RESEARCH AND DEVELOPMENT MANDATES

In response to the new congressional mandates discussed above, the Federal Aviation Administration (FAA) has developed an Aircraft Safety Research Program which includes a Fire Research Program. These programs are discussed below.

Aircraft Safety Research Program

In the two research areas of aviation maintenance and fire safety, the FAA Technical Center (FAATC) established in 1991 a multifaceted Aircraft Safety Research Program and an Aircraft Safety Research Plan (FAA, 1991). The Plan concentrates on identifying improvements to the aircraft and its systems, specifically in the areas of:

- structural inspection and repair,
- systems reliability,
- crew alerting and awareness,
- fault detection,
- failure prevention,
- crash energy absorption,

- fuel flammability, and
- materials fire resistance.

These improvements will be directed primarily to newly designed aircraft, since by the year 2000 much of the current jet fleet will have been replaced by aircraft that will contain more technologically advanced electrical and mechanical systems, and their controls will most likely be electronic rather than mechanical.

FAATC has established the Aircraft Safety Research Program's goal to be a 50 percent reduction, by the year 2000, in the current rate of fatalities due to aircraft accidents. Since the number of people using airplanes is expected to increase at an annual rate of 5 percent, a reduction of 50 percent in the number of fatalities over current rates by the year 2000 actually equates to a 62 percent reduction in the rate of fatalities.

While fatalities have occurred in accidents that did not involve a crash, such as in in-flight fires, by far the largest number of fatalities occur because of crashes that destroy or disable the airplane structure. Crashes are divided roughly into nonsurvivable and survivable categories. Nonsurvivable crashes are so violent (e.g., CFIT) that the impact trauma is expected to be fatal for all airplane occupants. Survivable crashes involve violence that is not as severe and at least some of the airplane occupants are expected to survive. The accident database used by the FAA in the Aircraft Safety Research Plan shows that, in the past, about half of aircraft accident fatalities have occurred in impact nonsurvivable crashes and half have occurred in impact survivable crashes.

In impact nonsurvivable crashes, lives can be saved only if the accident is prevented or made somehow survivable. In impact survivable accidents, improved post-crash survival provisions, in addition to accident prevention, can also be effective. Since about 50 percent of accident fatalities in the past have occurred in impact nonsurvivable accidents, the Plan points out that it is clearly not possible to achieve the Program's goal of reducing the fatality rate by 62 percent through implementation of post-crash survival measures alone, irrespective of how effective such measures may be. The FAA's goal therefore requires that accident prevention be a crucial part of the Plan. Hence the FAA established the Program's first-tier objective to be the prevention of aircraft systems failures that would cause an accident.

If an accident does occur and it is an impact survivable crash, improved post-crash impact protection provisions would be beneficial since, in the past, about 30 percent of all aircraft accident fatalities have been due to impact in impact survivable crashes. Therefore, a second objective is to improve passenger survivability due to impact in impact survivable crashes.

If an impact survivable crash does occur and passengers survive the impact, the major threat thereafter is fire fed by jet fuel. In the past, about 20 percent of all aircraft accident fatalities have been due to fire in impact survivable crashes. Therefore a third objective is to prevent fire in impact survivable crashes. If fire does occur, a fourth objective is to retard its spread into and through the passenger cabin in order to provide sufficient time for passengers to evacuate the burning aircraft before they are burned or become incapacitated by heat or toxic gases.

Fire Research Program

Included in the Aviation Safety Research Act of 1988 was a specific mandate for the FAA to establish research efforts "to assess the fire and smoke resistance of aircraft materials, to develop and improve fire and smoke containment systems for in-flight aircraft fires, and to develop advanced aircraft fuels with low flammability and technologies for containment of aircraft fuels for the purpose of minimizing post-crash fire hazards."

For many years the FAATC has made a substantial effort in research on fire safety. The Fire Safety Branch has supported the development of several new Federal Aviation Requirements regulations (e.g., more fire-resistant passenger seats, passenger cabin liners, cargo compartment liners, etc.), participated in NTSB investigations of aircraft accidents involving fire, and provided applied research and test support to the aircraft industry as well as other FAA organizations. To comply with the mandates of the new legislation to expand its fire-safety work to encompass more fundamental fire-safety research, the FAA has established a Fire Research Program and developed a comprehensive Fire Research Plan (FAA, 1993). The goal of the Fire Research Program is to eliminate fire as a cause of fatalities in aircraft accidents.

To implement the Program, the FAA created a separate Fire Research Branch and staffed it with experienced and knowledgeable personnel, some of whom had previously been associated with the Fire Safety Branch. The Fire Research Plan has identified the following fire research areas to be pursued:

- fire modeling,
- vulnerability analysis,
- fire-resistant materials,
- improved systems,
- advanced suppression, and
- fuel safety.

The technical products of the research will include:

- new fire-safety design tools,
- new technology safety products,
- more economical fire-suppression systems,
- ultra-fire-resistant materials,
- tailored fuel properties, and
- advanced fire-safety assessment technologies.

Equipment and personnel to support studies in fire modeling and fire-resistant materials were already available, so activities in these two areas were initiated in 1993.

FAA Interim Materials Development Criteria

Since there is at present a lack of knowledge concerning the relationships between a material's composition and how to relate its bench-scale fire-test performance to large-scale tests, the FAATC has proposed extremely conservative, preliminary materials fire performance guidelines as the initial screening criteria. However, as fire modeling and probabilistic risk assessment tools from other initiatives in the Fire Research Program improve over the course of the Program and more full-scale test data become available to better relate bench-scale fire test data to ignition and flame spread in aircraft, fire performance guidelines may be relaxed consistent with maintaining a totally fire-resistant aircraft cabin as demonstrated in full-scale performance tests.

The preliminary screening criteria proposed by the FAA parallel those established by the U.S. Navy for composite materials used in submarines (DeMarco, 1991; DOD, 1991). The interim guidelines adopt the Navy requirements for oxygen index, flame-spread index, combustion gas generation, smoke, and smoke toxicity, but extend the ignitability and heat release guidelines to require essentially noncombustible behavior and add a full test requirement.

• Ignitability

Objective: To reduce the propensity for ignition of cabin materials in a post-crash jet fuel fire.

Requirements: No piloted ignition when tested at 50 kW/m^2 irradiance in accordance with ASTM E-1354 incident heat fluxes of 50 ± 10 kW/m^2 (corresponding to an equilibrium surface temperature of about 650°C) are measured near the bottom and center of open doors in passenger aircraft exposed to external jet fuel fires. Organic polymeric materials with decomposition temperatures in this range are nonignitable when tested at 50 kW/m^2 irradiance (Kim et al., 1993).

• Heat Release

Objective: To delay cabin flashover in a post-crash jet fuel fire.

Requirements: Maximum heat release rate less than 50 kW/m^2 for 1.6-mm-thick materials tested in a vertical orientation at an irradiance level of 75 kW/m^2 in accordance with ASTME-1354. Heat fluxes measured above open doors in passenger aircraft exposed to external fuel fires are 75 ± 50 kW/m^2, depending on fire size and wind conditions (Brown, 1979; Hill et al., 1979; Quintiere et al., 1985). Correlation of time-to-flashover data from full-scale aircraft cabin fire tests with both $Q_c8/3$ and the flashover parameter $Q_c peak/t_{ign}$ indicate that the maximum heat release rate must be less than about 40–50 kW/m^2 to delay flashover of thermally stable (T_{decomp} ~ 650°C) cabin materials for 15 minutes in a post-crash fire scenario.

• Full-Scale Aircraft Cabin Fire Tests

Objectives: Performance criteria for fire-safe materials systems. Demonstrate survivable aircraft cabin conditions for 15 minutes in post-crash fuel fires.

Requirement: No flashover or incapacitation from combustion gases for at least 15 minutes in full-scale aircraft cabin fire tests by the FAA under quiescent wind conditions.

These fire performance guidelines will be met by synthesizing new materials; characterizing and modeling their thermochemical and thermophysical behavior in fires at the atomic, micro-, and macroscopic levels; and applying this knowledge to the design of new materials with better thermal stability, lower heat release, and tailored thermophysical properties. Supporting research in processing and collaboration with the materials and aircraft industries will help ensure that advanced fire-safe materials are also cost-effective, reliable, and serviceable so as to increase the fireworthiness and add value to future aircraft.

The need for fire-safe materials systems that are light-weight with better strength and serviceability and with lower installation costs in aircraft will be satisfied by polymers, ceramics, composites, and additives which are economical to synthesize and process and have minimum environmental impact. Research will be basic in nature and will focus on synthesis, characterization, modeling, and processing of new materials and materials combinations to improve the fire performance, increase the reliability, and reduce the cost of next-generation cabin materials. A description of what is needed in each of these research areas follows.

Synthesis

The synthesis effort will focus on developing materials and synthetic routes to materials that are environmentally sound and possess exceptional thermal and hydrolytic stability, high char yield/low pyrolysis fraction, low density, low heat release, and moderate temperature or thermal processing requirements. Applications include rigid and resilient foams, decorative films, textile fibers, adhesives, and fiber-reinforced composites of interest are novel materials, materials combinations, additives, blends, and composites. These include, but are not limited to, melt processable/soluble liquid-crystal polymers, self-reinforcing molecular composites, nanocomposites, nanophase materials, multiphase materials, organometallics, organic/inorganic hybrids and copolymers, low-cost polymer-precursor ceramic materials, inorganic resins, and high strength/low density ceramic resins and coatings. Low-cost nanoscale reinforcements (fibers, whiskers), which could impart fire-resistance through surface catalytic activity or improve specific mechanical performance of base resins to reduce fire load, are of interest.

Needed are novel halogen-free polymers, phosphorous-, silicon-, boron-, and sulfur-containing polymers—particularly in combination with characterization and modeling activities that seek to understand the relationship between backbone or pendant heteroatoms and the pyrolysis kinetics and char-forming tendency of organic materials. Synthesis and characterization of metal-containing polymers and additives that mimic the high efficiency and synergistic fire-retardant activity of antimony compounds is needed. Renewable, biodegradable, fire-resistant, low-density materials are required as are concepts for polymers that self-extinguish by decomposing into gas-phase fire suppressants. A "master polymer resin" may be developed that would combine exceptional fire-resistance mechanical properties and ease of processibility so as to find broad use as aircraft cabin materials and in structural composites at reasonable installed cost.

Mechanistic studies to understand and discover novel organic, inorganic, and semi-inorganic polymerization reactions are important. Thermally stable, crosslinking reactions for structural thermoset resins that do not generate volatile products (e.g., cyanate, acetylenic, Michael addition) should be studied mechanistically, and new crosslinking reactions for organic and inorganic resins should be discovered or adapted to fire-safe polymers. The synthesis effort is expected to yield:

- Fire-safe thermoset and thermoplastic engineering polymers suitable for use in cabin interior and structural applications including molded parts, adhesives, films, rigid foams, coatings, and composite matrices. Material should be low in cost; easily fabricated at moderate temperatures on existing or available processing equipment; generate no volatile products during cure; possess high strength, toughness, and hygrothermal durability; and be environmentally benign during fabrication, use, and disposal.
- Fire-safe elastomer suitable for use in resilient foam applications (seat cushions, pillows), adhesives, and sealants. Material should be low in cost; easily processible at moderate temperatures on existing or available equipment; have high elongation, good recovery, abrasion and tear resistance, and hygrothermal durability; and be environmentally benign during fabrication, use, and disposal.
- Fire-safe fibers for seating upholstery, tapestry, blankets, and carpeting application. Material should be low in cost; processable into fibers and yarns on existing textile fiber spinning equipment; have good mechanical properties, colorability, and hygrothermal durability; and be environmentally benign during fabrication, use, and disposal.

Characterization

Fundamental new test methods that are less expensive and can more reliably predict or simulate fire behavior of low heat release materials in service are needed. Also of interest are innovative analytical techniques and combinations of techniques that provide new insight and fundamental information about the thermochemical and thermophysical processes of solid-state thermal decomposition, thermo-oxidation, char formation, pyrolysis, gasification, nonflaming and flaming combustion in support of molecular modeling and mechanistic materials fire models. A better understanding of fire behavior, including radiant and convective energy transport at burning surfaces, flame chemistry, smoke and soot generation and their radiant properties, are necessary for development of engineering materials models and for application of these models to bench scale reaction-to-fire tests such as the Cone Calorimeter. The effect of flame-retardant additives on the heat release, mass-loss rate, heat of gasification, char, and smoke production need to be measured and correlated with full-scale fire behavior of organic polymeric materials at elevated heat fluxes. Also needed is an understanding of the effect of physical combinations of materials (e.g., laminates, coatings, fire blocking-layers, etc.) representing real component construction on fire-related properties. Rheology, thermal analysis, and mechanics studies to determine the utility of ultra-low-density materials, polymers, polymer-additive systems, blends, networks, composites, fibers, and elastomers with demonstrated fire safety or fire-safety potential are also needed. Research will seek to develop understanding of the underlying principles connecting physical, mechanical, and chemical phenomena. The characterization effort is expected to yield:

- analytical methodologies for characterizing combustion of low heat release materials,
- bench-scale reaction-to-fire test methods for low heat release materials, and
- thermal properties of chars and materials at fire temperatures.

Modeling

Computational and theoretical modeling of the chemical and mechanical behavior of materials exposed to fire and severe thermal environments is needed. Atomistic computational modeling of homogeneous materials using molecular dynamics simulations is needed to study solid-state thermal decomposition, thermo-oxidation, char formation, and gasification (fuel production) rate in relation to inter-and intramolecular bond strength(s), bond strength distributions, molecular mobility, diffusion of gases and small molecules, and the role of morphological features on these processes. Atomistic computational modeling of thermally induced reactions of metal/polymer mixtures will help understand the mechanistic processes responsible for the anomalous high efficiency and synergistic activity of antimony-oxide flame retardants and help design replacements for these potentially toxic heavy metal compounds. Computational studies of surface catalytic effects on thermal and thermo-oxidative degradation and char formation processes at internal material interfaces must be performed to assess and design effective additives, blends, and surface treatments for fibers and particulate additives in an effort to explore the limits of reduced flammability using existing materials. Theoretical analytical models that seek to describe and predict reaction kinetics, phase behavior, and miscibility of fire-resistant polymer blends and fire-retardant additives are needed.

Determining the fire hazard of polymers and polymer composite materials in airframes, skins, and cabins will require accurate engineering models which capture the essential physics, chemistry, and mechanical impact of heat transfer, gasification, heat release, flame spread, smoke generation, and char formation for use in field fire models of post-crash scenarios. Mechanistic analytic models for thermal degradation, pyrolysis, and char formation are necessary to gain a physical understanding of the relationship between these quantities and to connect the molecular modeling with engineering fire models.

Sophisticated thermostructural models must be developed for fire-exposed composites which can account for strength degradation. Theoretical mechanics models of time-and temperature-dependent buckling and creep of viscoelastic polymers and anisotropic polymer composites are necessary to assess the structural capability of fire-exposed airframes, empennage, and secondary structures such as floor beams. Theoretical mechanical models will be developed to provide guidance for toughening and durability modification of typically brittle, high-temperature polymers used as resins and adhesives, and to identify novel high-strength micro-and macrostructures for ultra-lightweight foams and cores. The modeling effort is expected to yield:

- validated computational methodologies for molecular design of fire-resistant polymers;
- mechanistic models for pyrolysis and char formation of aircraft materials;
- engineering models for ignition, combustion, heat release, heat transfer, and flame spread of materials for use in risk assessment (structural response codes and fire field models); and
- theoretical and engineering models for load-bearing capability of polymers and composites during and after fire exposure.

Processing

Fabrication processes for advanced fire-safe materials must be fully reproducible, verifiable, and able to maintain tight tolerances in order to replace existing materials in passenger aircraft. Research on processing to reliably provide novel materials with uniformly improved properties at lower cost is of interest. On-line remote process monitoring for continuous processing via reactive extrusion and pultrusion are of interest particularly when coupled with chemometric or neural network techniques for intelligent processing. Relationships among processing, fire performance, and mechanical performance are of interest for reactive extrusions and processing-generated microstructures. Chemorheology of fire-safe thermoset reactions, low-temperature processing routes to high-temperature materials, electromagnetic processing (ultrasonics, microwave, e-beam, etc.), rheology of ternary blends, process sensors, recycling concepts for noncombustible materials are of interest. Also of interest are novel processing-generated microstructures with superior fire resistance and innovative processing routes to high-strength, ultra-low-density core materials and materials systems. The modeling effort is expected to yield:

- on-line process monitoring and control technology for high-volume, economical, tight tolerance production of polymers and composites;
- low-temperature routes to high-temperature-capable materials; and
- novel processing-generated microstructures and ultra-low-density core materials.

Formal collaboration among materials researchers and fire scientists in academia, government, and industry is anticipated. Installation of new materials in aircraft will require close cooperation among researchers, materials manufacturers, and the aircraft industries to develop a supplier base, manufacturing technology, and cost and performance requirements for new materials. The FAA anticipates sponsoring an industrial research program in parallel to the academic research program as part of the fire-safe materials effort to encourage the development of precompetitive industrial technology for fire-safe materials and to leverage core technologies and commercial production capabilities of key chemical and materials producers.

REFERENCES

Brown, L.J. 1979. Cabin Hazards From a Large External Fuel Fire Adjacent to an Aircraft Fuselage. FAA-RD-79-65. Atlantic City, N.J.: Federal Aviation Administration Technical Center.

DeMarco, R.A. 1991. Composite applications at sea: fire-related issues. Proceedings of the 36th International SAMPE Symposium 36(2):1928–1937.

DOD (U.S. Department of Defense). 1991. Fire and Toxicity Test Methods and Qualification Procedure for Composite Material Systems Used in Hull, Machinery, and Structural Applications Inside Naval Submarines. MIL-STD-2031(SH). Washington, D.C.: U.S. Department of Defense.

FAA (Federal Aviation Administration). 1991. Air Safety Research Plan . Atlantic City, N.J.: FAA Technical Center.

FAA (Federal Aviation Administration). 1993. Fire Research Plan. Atlantic City, N.J.: FAA Technical Center.

Hill, R.G., G.R. Johnson, and C.P. Sarkos. 1979. Postcrash Fuel Fire Hazard Measurements in a Wide-Body Aircraft Cabin. FAA-NA-79-42. Atlantic City, N.J.: Federal Aviation Administration Technical Center.

Kim, P.K., P. Pierini, and R. Wessling. 1993. Thermal and flammability properties of poly (p-phenyleneben-zobisoxazole). Journal of Fire Sciences 112(4):296–307.

Quintiere, J.G., V. Babrauskas, L. Cooper, M. Harkleroad, K. Steckler, and A. Tewarson. 1985. The Role of Aircraft Panel Materials in Cabin and Their Properties. DOT/FAA/CT-84/30. Atlantic City, N.J.: Federal Aviation Administration Technical Center.

Appendix C

Current Fire-Modeling Capabilities

As described in Chapter 3, computer models of ignition and flame spread are used to improve the understanding of fire scenarios, to evaluate small-scale material test results and provide an estimation of full-scale performance, and to aid in the development of fire performance goals. Fire-hazard assessment models are used to characterize the hazards associated with a fire event. These models use the results of the two general types of fire models—zone models and field models—along with submodels to account for the effects of detection and extinguishing systems and toxicity. This appendix describes current modeling capabilities.

ZONE MODELS

Zone models are often applied to buildings composed of several adjoining compartments, where each compartment is subdivided into control volumes or zones. All quantities of interest are uniform within each zone; conservation of mass, energy, and momentum is applied to each zone using algebraic representations or ordinary differential equations. The advantages of zone models are the low memory requirements, speed, ability to represent large structures, a priori assumptions, structured output, and ease of use. The need for a priori assumptions and the inaccuracies that may result can also be viewed as a disadvantage.

There are a number of zone models in use in the United States. One that is in wide use is the CFAST model developed by the National Institute of Standards and Technology (Jones, 1985; Jones and Peacock, 1988; Jones and Forney, 1990). This model is also used in fire reconstruction, particularly in litigation procedures. Other well-known zone models are BRI2 (Tanaka, 1977, 1978, 1980) developed in Japan and used extensively in risk analysis at Factory Mutual Research Corporation.

To gain an understanding of the state of the art of zone modeling, the capabilities of the CFAST model are:

- multiple compartments (currently 15);
- multiple fires (currently 16)—can be specified with "other" objects;
- vitiated or free-burn chemistry in the lower layer, the upper layer, or in the vent flow;
- correct chemistry—consistent production and transport of species;
- generalized species (10) transport;
- four-wall and two-layer radiation to be extended for the pyrolysis model;
- four-wall conductive heat transfer through multilayered walls, ceilings, and floors in each compartment;
- convective and radiative heat transfer applied to both inside and outside boundaries;
- wind effects—American Society of Heating, Refrigeration and Air-Conditioning Engineers formula for wind with the National Oceanographic and Atmospheric Administration integral for lapse rate of the standard atmosphere;
- fire plume and entrainment in vent flow (doors and windows only)—fire plume is split into entrainment in the lower and upper layer;
- three-dimensional specification of the location of the fire and nonuniform heat loss through boundaries;
- generalized horizontal vent flow (doors, windows, etc.)—up to three neutral planes; mixing between the upper and lower layers; vertical flow (through holes in ceilings and floors);
- separate internal and external ambient (elevation, temperature, and pressure specification);
- hydrochloric acid deposition;
- mechanical ventilation—complex building structure: 5 fans, 44 ducts, 3-way joints; vertical ducts interact with both layers; and
- heat transfer (conductive) through barriers.

The method by which these types of models achieve the speed and low memory requirements that make them attractive is through the use of a priori assumptions. These assumptions are input parameters to the model and are derived from empirical data. One of the concerns that needs immediate attention is to increase the confidence level and reliability of model predictions. These goals can be achieved in two ways. First, close interaction is needed between the modeler and experimentalist to identify the empirical data required by the model and to then create materials property databases that contain the needed information. Second, since small-scale experimental data are being used to "scale up" and predict actual fire behavior in models, it is imperative that modeling predictions be verified in large-scale experiments. For example, CFAST model predictions have been validated with experimental data from the Navy's fire research platform ex-U.S.S. Shadwell (Williams and Carhart, 1992).

FIELD MODELS

There are a number of field models available (Yang et al., 1984; Kou et al., 1986). A two-dimensional finite-difference field model of aircraft fires was developed to predict the movement of hot gases and smoke, as well as temperature and smoke concentration levels in the seating area of an aircraft cabin (Yang et al., 1984; Kou et al., 1986). Additional work included the development of a two-dimensional model of transient cooling by natural convection (Nicolette et al., 1985). This model utilized a fully transient semi-implicit upwind-differencing scheme with a global pressure correction. Baum and Rehm (1978, 1982a, b, 1984) have developed several field models for prediction of fires using time-dependent inviscid Boussinesq equations to simulate three-dimensional buoyant convection and smoke aerosol coagulation.

There have been several field modeling projects to develop a computer model as a low-cost alternative to predict the spread of fire and smoke in enclosed spaces on naval vessels (Nies, 1986; Raycraft, 1987; Hauck, 1988). The similarity of the enclosed spaces of naval vessels to an aircraft interior makes these types of models valuable in evaluating the effectiveness of suppression systems and new designs in the prevention and control of fires in aircraft.

Field modeling requires a large, fast computer with significantly more memory than is required in zone modeling. The accuracy of the solution depends on reducing the size of the control volumes, thus increasing the number of individual cells and the computing expense.

Fire-Hazard Assessment Models

Fire-hazard assessment models include both zone and field models for compartment fires, with submodels for fire endurance, activation of thermal detectors or sprinkler systems, generation of toxic gases, evacuation, and survival models.

One of the early room fire models, HARVARD V, was developed in the early 1980s (Emmons, 1981; Mitler and Emmons, 1981). With this model, the user specifies room characteristics, technical information on objects contained in the room, and where the fire starts. The program calculates the fire growth, fire plume, the accumulated hot layer at the ceiling, and the outflow of hot gases and inflow of air after the smoke layer reaches the soffit. The program also calculates the radiation from the flame and hot layer to all of the objects. As each object reaches ignition temperature, the program ignites a new fire and new plume and models the additional hot gas going to the upper layer. As the flames and hot layer grow, radiation to the burning objects controls the rate of fire growth over their surface. When all the objects ignite, flashover has occurred. As the hot layer descends and envelops a burning object, the program calculates the reduction of airflow and the fire slows down. The program keeps track of the total mass and indicates when the fire of each object goes out. It also calculates the time at which a smoke detector in the room will sound its alarm.

Two fire-hazard assessment models currently in use are HAZARD 1 (Bukowski et al., 1991) and FPEtool (Nelson, 1990). According to the authors, HAZARD 1 will calculate:

- the production of energy and mass (smoke and gases) by one or more burning objects in one room, based on small-or large-scale measurements;
- the buoyancy-driven ventilation, as well as forced flow, of this energy and mass through a series of user-specified rooms and connections (doors, windows, cracks, holes in ceiling or floor);
- the resulting temperatures, smoke optical densities, and gas concentrations after accounting for heat transfer to surfaces and dilution by mixing with clean air;
- the evacuation of a user-specified set of occupants accounting for delays in notification, decision making, behavioral interactions, and inherent capabilities; and
- the impact of the exposure of these occupants to the predicted room environments as they move through the building; and the time, location, and cause of each incapacitation or fatality.

This model requires detailed knowledge of the fire scenario, including the geometry of the room(s), the location of the items that are burning, the combustion properties of those items, and also information on the exposed occupants (i.e., their initial location and characteristics such as age, and whether they have any disabilities or small children to assist). Therefore, the users of this model need to be familiar with fire physics and understand the limitations of the model.

FPEtool is also a hazard assessment computer model (Nelson, 1990), but is less mathematically rigorous than HAZARD 1 and therefore takes less time to run. It will estimate ignition of exposed objects, smoke flows, gas concentrations and toxicity, pressures on a door from the fire and wind, actuation of detectors and sprinklers, and egress time of occupants.

Table C-1 describes some limits inherent to HAZARD 1, FPEtool, and HARVARD V and which probably bracket the majority of limitations inherent within two-zone models. However, there is an incomplete understanding of the physical phenomenon involved in fires. Each limitation results in a computer code that may deviate from the correct representation of the fire physics that could introduce errors into

TABLE C-1 Characteristics and Limitations of Fire-Hazard Assessment Models

Prediction	HARVARD V	HAZARD 1	FPEtool
Ignition of primary fuel	Specified by user	Specified by user	Specified by user
Toxic gas release	Calculated by program after rate of burning calculations	Specified by user	Specified by user
Pyrolysis enhancement from smoke-layer radiation	Calculated by program as is pyrolysis enhancement from secondary fuel fire radiation	Specified by user	Specified by user
Fire growth	Calculated by program	Specified by user	Specified by user
Ignition of secondary fuel	Calculated by program for all fuels in a compartment as user specified	Planned	Calculated by program
Flames burning in door jets	No	Yes, but predicted from a limited set of experimental data	No
Plume entrainment	Correlated to a limited set of experimental data	Correlated to a limited set of experimental data, not based on fundamental fluid dynamics	Correlated to a limited set of experimental data, not based on fundamental fluid dynamics
Radiation transfer from smoke	Calculated by program from hot smoke layer and flames to all specified items in compartment chosen by user	Considers reasonable detail to floors, walls, and ceilings	Considered in correlational approach for walls, reasonable detail to the floor
Flashover	Program calculates when all items ignite and when room reaches 500°C	Oxygen starvation of fire is considered, otherwise same as preflashover treatment	Calculated distinctly from preflashover burning. No plume calculations.
Heat transfer from one room to another	No	Planned	No
Failure of wall or ceiling barrier	No	Planned	No
Activation of sprinkler/heat detector remote from fire	No	No	No
Unconsciousness/death based on toxic gases	No	Yes, prediction is based on N-gas model of four gases: CO, CO_2, O_2, and HCN.	Yes, prediction is based on N-gas model of four gases: CO, CO_2, O_2, and HCN.
Death based on atmospheric temperature	No	Yes, assumes instantaneous death when room temperature reaches 100°C.	Yes, based on energy absorption by the body; data from literature

Source: HAZARD 1 and FPEtool—S. Deal and R. Peacock to B.C. Levin, personal communication, January, 1995.

the calculations. A further complication arises when limitations inherited from the numerical representation of one physical phenomenon compound, potentiate, or negate limitations inherited from the numerical representation of a second physical phenomenon.

None of the fire-hazard models was designed to be applicable to aircraft interior fires since they model rooms rather than the cylindrical geometry of an aircraft interior. The models assume uniform smoke filling the upper layer with no time lag to compensate for the length of the corridor-like space and point-source fires of an aircraft. Also, the behavior and evacuation of individuals modeled in HAZARD 1 is from one-and two-family residences which is a very different scenario from the behavior and evacuation of people from an airplane. As in all fire models, the definition of the fire will depend on the materials that are burning. New materials will need to be characterized to obtain the data necessary to input into these models. Thus, their use to model an interior fire on an aircraft will take them outside their domain of applicability.

REFERENCES

Baum, H.R., and R.G. Rehm. 1978. The equations of motion for thermally driven, buoyant flows. Journal of Research of the National Bureau of Standards 83(3):297–308.

Baum, H.R., and R.G. Rehm. 1982a. Computation of fire induced flow and smoke coagulation. Pp. 921–931 in Nineteenth Symposium (International) of Combustion. Pittsburgh, Pa.: The Combustion Institute.

Baum, H.R., and R.G. Rehm. 1982b. Natural Computation of Large Scale Fire-Induced Flows. Paper presented at the Eighth International Conference on Numerical Methods in Fluid Dynamics, Aachen, West Germany, June 28–July 2.

Baum, H.R., and R.G. Rehm. 1984. Calculations of three dimensional buoyant plumes in enclosures. Combustion Science and Technology 40:55–77.

Bukowski, R.W., R.D. Peacock, W.W. Jones, and C.L. Forney. 1991. Technical Reference Guide for HAZARD 1 Fire Assessment Method, Version 1.1, Vol. 2. National Institute of Standards and Technology Handbook 146/II. Gaithersburg, Md.: National Institute of Standards and Technology.

Emmons, H.W. 1981. The Calculation of a Fire in a Large Building. ASME Paper 81-HT-2 for meeting of the American Society of Mechanical Engineers, August 2–5, 1981.

Hauck, R.R. 1988. Numerical Field Model Simulation of Full-Scale Fire Tests in a Closed Spherical/Cylindrical Vessel with Internal Ventilation. Unpublished master's and mechanical engineer's thesis, Naval Postgraduate School, Monterey, California.

Jones, W.W. 1985. A multicompartment model for the spread of fire, smoke and toxic gases. Fire Safety Journal 9:55.

Jones, W.W., and G.P. Forney. 1990. A Programmer's Reference Manual for CFAST, The Unified Model of Fire Growth and Smoke Transport. NIST Note 1283. Gaithersburg, Md.: National Institute of Standards and Technology.

Jones, W.W., and R.D. Peacock. 1988. Refinement and experimental verification of a model for fire growth and smoke transport. Pp. 897–906 in the Proceedings of the Second International Symposium on Fire Safety Science, T. Wakamatsu, Y. Hasemi, A. Sekizawa, P.G. Seeger, P.J. Pagni, and C.E. Grant, eds. New York: Hemisphere Publishing Corp.

Kou, H.S., K.T. Yang, and J.R. Lloyd. 1986. Turbulent buoyant flow and pressure variations around an aircraft fuselage in a cross wind near the ground. Pp. 173–184 in Fire Safety Science, Proceedings of the First International Symposium, C.E. Grant and P.J. Pagni, eds. New York: Hemisphere Publishing Corp.

Mitler, H., and H. Emmons. 1981. Documentation for CFS V, the Fifth Harvard Computer Fire Code. NBS-GCR-81-344. Gaithersburg, Md.: National Institute of Standards and Technology .

Nelson, H.E. 1990. FPEtool: Fire Protection Engineering Tools for Hazard Estimation. NISTIR 4380. Gaithersburg, Md.: National Institute of Standards and Technology.

Nicolette, V.F., K.T. Yang, and J.R. Lloyd. 1985. Transient cooling by natural convection in a two-dimensional square enclosure. International Journal of Heat Transfer 28(9):1721–1732.

Nies, G.F. 1986. Numerical Field Model Simulation of Full Scale Tests in a Closed Vessel. Unpublished master's and mechanical engineer's thesis, Naval Postgraduate School, Monterey, California.

Raycraft, J.K. 1987. Numerical Field Model Simulation of Full Scale Fire Tests in a Closed Spherical/Cylindrical Vessel. Unpublished master's and mechanical engineer's thesis, Naval Postgraduate School, Monterey, California.

Tanaka, T. 1977. A Mathematical Model of a Compartment Fire. Report 70. Ibaraki-ken, Japan: Building Research Institute.

Tanaka, T. 1978. A Mathematical Model of a Compartment Fire. Report 79. Ibaraki-ken, Japan: Building Research Institute.

Tanaka, T. 1980. A Mathematical Model of a Compartment Fire. Report 84. Ibaraki-ken, Japan: Building Research Institute.

Williams, F.W., and H.W. Carhart. 1992. The Ex-Shadwell—Full Scale Fire Research and Test Ship. NRL Memorandum Report 6074. Washington, D.C.: Naval Research Laboratory.

Yang, K.T., J.R. Lloyd, A.M. Kanury, and K. Satoh. 1984. Modeling of turbulent buoyant flows in aircraft cabins. Combustion Science and Technology 39:107–118.

Appendix D

Toxicity Modeling Testing

The toxicity of the combustion products from new fire-safe materials will need to be determined. The terms "fire safe" or "fire resistant" are not the same as noncombustible. Unless new material is truly noncombustible, some thermal decomposition will occur when the material is exposed to fire conditions. Both the toxic gases and the irritants that are present in all smoke need to be considered as potential dangers. The toxic products can cause both acute and delayed toxicological effects. It is the acute and extremely short-term effects that prevent escape from an aircraft, causing faulty judgment, incapacitation, and death. The irritants in smoke can also interfere with passengers' ability to escape by causing severe coughing and choking and by preventing them from keeping their eyes open long enough to find the exits. In addition, the delayed effects, such as tissue or organ injury, mutagenicity, carcinogenicity, and teratogenicity need to be characterized since they may ultimately lead to permanent disability and post-exposure deaths.

TOXICITY MODELS

There is currently no national standard test method for evaluating smoke toxicity in the United States, although the American Society for Testing Materials is close to approving the Radiant Heat Method ASTM E-1678 (Babrauskas et el., 1991a; Levin, 1992a, b). The International Standards Organization, Technical Committee 92, Subcommittee 3 (ISO/TC92/SC3) on Toxic Hazards in Fire has approved an international standard for combustion toxicity. This standard (ISO/DIS 13344 "Determination of the Lethal Toxic Potency of Fire Effluents") describes the mathematical models available for predicting the toxic potency of fire atmospheres based on the toxicological interactions of the main combustion gases present (Hartzell, 1994; Levin et al., 1995; Purser, 1995). Rather than designate a specific combustion system, investigators have the flexibility of designing or choosing a system that will simulate conditions relevant to their fire scenario. One of the models, the N-gas model (included in ASTM E-1678 and ISO/DIS 13344) is an empirical mathematical relationship containing six gases—carbon monoxide (CO), carbon dioxide (CO_2), low oxygen (O_2), hydrogen cyanide (HCN), hydrogen chloride (HCI), and hydrogen bromide (HBr)—and is shown in Equation (1) (Levin et al., 1987a, b, 1995).

$$\text{N-Gas Value} = \frac{m[CO]}{[CO_2]-b} + \frac{[HCN]}{LC_{50}HCN} + \frac{21-[O_2]}{21-LC_{50}O_2} + \frac{[HCl]}{LC_{50}HCl} + \frac{[HBr]}{LC_{50}HBr} \quad (1)$$

The numbers in parentheses indicate the time-integrated average atmospheric concentrations during a 30-minute (or other relevant time) exposure period ([ppm • min]/min) or for O_2, the units are ([percent • min]/min). Although concentrations of CO_2 generated in fires are not lethal, there is a synergistic effect between CO_2 and each of the other tested gases. But the effect of CO_2 can only be included in the model once. Therefore, it is included with the CO factor. Empirically, it was found that as the concentration of CO_2 increases (up to 5 percent), the toxicity of CO increases. In the presence of amounts of CO_2 greater than 5 percent, the toxicity of CO starts to revert back toward the toxicity of CO by itself. The terms m and b in Equation (1) define this synergistic interaction. For example, in 30-minute exposures, $m = -18$ and $b = 122{,}000$ if the CO_2 concentrations are 5 percent or less. For 30-minute studies in which the CO_2 concentrations are above 5 percent, $m = 23$ and $b = -38{,}600$. The LC_{50} value of HCN is 200 ppm for 30-minute exposures or 150 ppm for 30-minute exposures plus the post-exposure observation period. The 30-minute LC_{50} of O_2 is 5.4 percent which is subtracted from the normal concentration of O_2 in air (i.e., 21 percent). The LC_{50} value of HC1 and HBr for 30-minute exposures plus post-exposures times is 3,700 ppm (Hartzell et al., 1990) and 3,000 ppm (W.G., Switzer, Southwest Research Institute. personal communication, 1995), respectively.

A new N-gas model was needed to move from the six-gas model (containing CO, CO_2, HCN, O_2 concentrations. HCI, and HBr) to a seven-gas model which also includes NO_2 (Levin, 1994; Levin et al., 1995). NO_2 primarily produces pulmonary edema and post-exposure deaths. In binary gas studies, NO2 increased the toxicity of all the tested within-exposure toxic gases except HCN with which an antagonistic effect was observed. The reverse was also seen; that is, the post-exposure effects of NO_2 were more toxic if the animals had simultaneously been exposed to binary mixtures plus NO_2, plus CO, CO_2, or O_2.

The new seven-gas model including NO_2 is presented in Equation (2).

$$\text{N-Gas Value} = \frac{m[CO]}{[CO_2] - b} + \frac{21 - [O_2]}{21 - LC_{50}(O_2)} + \left(\frac{[HCN]}{LC_{50}(HCN)} \times \frac{0.4\,[NO_2]}{LC_{50}(NO_2)}\right) + 0.4\left(\frac{[NO_2]}{LC_{50}(NO_2)}\right) + \frac{[HCl]}{LC_{50}(HCl)} + \frac{[HBr]}{LC_{50}(HBr)} \quad (2)$$

where terms are defined as for Equation (1). The LC_{50} for NO_2 is 200 ppm following 30-minute exposures. Either the six-gas (Equation 1) or the seven-gas (Equation 2) model can be used to predict deaths that will occur only during the smoke exposure or those that will occur during and following the exposure. The seven-gas model is used if NO_2 predict the lethal toxicity of atmospheres that do not include NO_2, Equation (1) is used.

The N-gas model has been developed into an N-gas method. This method reduces the time necessary to evaluate a material and the number of test animals needed for the toxic potency determination. It also indicates whether the toxicity is usual (i.e., the toxicity can be explained by the measured gases) or is unusual (additional gases are needed to explain the toxicity). To measure the toxic potency of a given material with this N-gas method, a sample is combusted under the conditions of concern, and the concentration of gases considered in the model are measured. Based on the results of the chemical analytical tests and the knowledge of the interactions of the measured gases, an approximate LC_{50} value is predicted. In just two additional tests, six rats are exposed to the smoke from a material sample size estimated to produce an atmosphere equivalent to the approximate LC_{50} level (this can be for exposure or exposure plus post-exposure). Since the concentration-response curves for animal lethalities from smoke are very steep, it is assumed that if some percentage (not 0 or 100 percent) of animals die, the experimental loading based on the predicted LC_{50} value is close to the actual LC_{50}. No deaths may indicate an antagonistic interaction of the combustion gases. The deaths of all of the animals may indicate the presence of unknown toxicants or other adverse factors. If more accuracy is needed, then a detailed LC_{50} can be determined.

Results using the N-gas method have shown the good predictability of this approach,[1] Validation studies looking at a series of materials and products under conditions ranging from laboratory bench-scale to full-scale room burns indicated that, in all cases, the six-gas model was able to predict deaths correctly (Braun et al., 1988, 1990. 1991; Babrauskas et al., 1990, 1991a, b). The seven-gas model works when the animals are exposed to various concentrations and combinations of the tested gases (Levin et al., 1995). Studies need to be done to ensure that the seven-gas model predicts the outcome when nitrogen-containing materials and products are thermally decomposed.

TOXICITY TESTING

Toxicity screening tests for both acute and delayed effects are needed to evaluate the combustion products, including irritant gases of any newly proposed aircraft interior materials and products. It is imperative that the materials and products be tested under experimental conditions that simulate the realistic fire scenarios of concern in aircraft interiors as described earlier in this appendix. Tests should be simple, rapid, inexpensive, use the least amount of sample possible (since, in many cases, only small amounts of the developmental material may be available), use a minimum number of test animals, and have a definitive toxicological endpoint for comparison with other material candidates. While faulty judgment and incapacitation of passengers in an aircraft fire are significant causes of worry since they can prevent escape and cause death, they are complex endpoints that cannot be directly measured. Death of experimental animals (e.g., rats) is a more definitive and easily determined endpoint and can be used to compare the relative toxicities of alternative materials. Using lethality as the sole endpoint assumes that materials with greater toxicity based on a lethality endpoint will also cause more severe incapacitation and impairment. The number of experimental animals can be significantly reduced by utilizing one of the predictive mathematical models developed for combustion toxicology (Hartzell, 1994; Levin et al., 1995; Purser, 1995).

Many test methods for the determination of the acute toxicity of combustion products from materials and products have been developed over the last two decades and continue to be developed and improved (Kaplan et al., 1983; Norris, 1988; Caldwell and Alarie, 1991; Levin, 1992a, b). Two methods that are well known and have been used to generate much toxicity data are University of Pittsburgh I, a flow-through smoke toxicity method (Alarie and Anderson. 1979, 1981) and the closed-system cup furnace smoke toxicity method developed at the National Institute of Standards and Technology (NIST) (Levin et al., 1982, 1991). More recently,

[1] The LC50 values given for use in equations (1) and (2) are dependent on the test protocol, on the source of test animals, and on the rat strain. It is important to verify the above values whenever different conditions prevail and if necessary, to determine the values that would be applicable under the new conditions.

the University of Pittsburgh II radiant furnace method (Caldwell and Alarie, 1991), a radiant furnace smoke toxicity protocol (Babrauskas et al., 1991a, Levin, 1992a, b) by NIST and Southwest Research Institute, and the National Institute of Building Sciences (NIBS) toxic hazard test method (Norris, 1988; Roux, 1988) have been developed.

The NIST radiant test and the NIBS toxic hazard test use the same apparatus—consisting of a radiant furnace, a chemical analysis system, and an animal exposure chamber—although the methods have different toxicological endpoints. The NIST method uses an approximate LC_{50}, based on the *mass* of material needed to cause lethality in 50 percent of the test animals during a 30-minute exposure or the 30-minute exposure plus a 14-day post-exposure period, as the determinant of toxicity. The number of animals needed to run the test is substantially reduced by first estimating the LC_{50} using the N-gas model as described above. The toxicological endpoint of the NIBS toxic hazard test is the IT_{50}, the irradiation *time* that is required to kill 50 percent of the animals within a 30-minute exposure or during a 14-day post-exposure time. The actual results of the NIBS test with 20 materials indicated that the test animals died in very short periods of time, and the test was unable to discriminate definitively among materials. These test results indicate that the use of mass was a better discriminator among materials and thus was a better indicator of acute toxicity than time (i.e., the smaller the mass necessary for an LC_{50}, the more toxic the material).

In their current form, both the NIST and NIBS radiant test methods are unsuitable for the purpose of acute toxicity testing of materials for aircraft interiors because they are designed to simulate a post-flashover scenario which is not relevant to the fires of concern in aircraft interiors. The premise for simulating a post-flashover fire is that most people that die from inhalation of toxic gases in *residential* fires are affected in areas away from the room of fire origin. Smoke and toxic gases are more likely to reach these distant areas following flashover. In *aircraft* interior fires, however, tests need to be designed to specifically address the relevant scenarios discussed in Chapter 3. In addition, tests need to be designed to address the issues of delayed effects such as tissue or organ injury, mutagenicity, carcinogenicity, and teratogenicity.

REFERENCES

Alarie, Y.C., and R.C. Anderson. 1979. Toxicologic and acute lethal hazard evaluation of thermal decomposition products of synthetic and natural polymers. Toxicology & Applied Pharmacology 51:341–362.

Alarie, Y.C., and R.C. Anderson. 1981. Toxicologic classification of thermal decomposition products of synthetic and natural polymers. Toxicology & Applied Pharmacology 57:181–188.

Babrauskas, V., R.H. Harris, Jr., E. Braun, B.C. Levin, M. Paabo, and R.G. Gann. 1990. Large-scale validation of bench-scale fire toxicity tests. Pp. 3–12 in Interflam '90 Fire Safety, C.A. Franks, ed. London: Interscience Communications Ltd.

Babrauskas, V., R.H. Hams, Jr., E. Braun, B.C. Levin, M. Paabo, and R.G. Gann. 1991 a. Large-scale validation of bench-scale fire toxicity tests . Journal of Fire Sciences 9:125–148.

Babrauskas, V., R.H. Harris, Jr., E. Braun, B.C. Levin, M. Paabo, and R.G. Gann. 1991b. The Role of Bench-Scale Test Data in Assessing Real-Scale Fire Toxicity. NIST Technical Note 1284. Gaithersburg, Md.: National Institute of Standards and Technology.

Braun, E., B.C. Levin, M. Paabo, J.L. Gurman, H.M. Clark, and M.F. Yoklavich. 1988. Large-Scale Compartment Fire Toxicity Study: Comparison With Small-Scale Toxicity Test Results. NBSIR 88-3764. Gaithersburg, Md.: National Institute of Standards and Technology.

Braun, E., S. Davis, J.H. Klote, B.C. Levin, and M. Paabo. 1990. Assessment of the Fire Performance of School Bus Interior Components. NISTIR 4347. Gaithersburg, Md.: National Institute of Standards and Technology.

Braun, E., J.H. Klote, S. Davis, B.C. Levin, M. Paabo, and R.G. Gann. 1991. An assessment methodology for the fire performance of school bus interior components. Pp. 855–864 in Fire Safety Science—Proceedings of the Third International Symposium , G. Cox and B. Langford, eds. New York: Elsevier.

Caldwell, D.J., and Y.C. Alarie. 1991. A method to determine the potential toxicity of smoke from burning polymers: III. Comparison of synthetic polymers to Douglas Fir using the UPitt II Flaming Combustion/Toxicity of Smoke Apparatus. Journal of Fire Sciences 9:470–518.

Hartzell, G.E. 1994. Smoke toxicity and toxic hazards in fires. ASTM Standardization News 22:38–44.

Hartzell, G.E., A.F. Grand, and W.G. Switzer. 1990. Toxicity of smoke containing HCI. Pp 12–20 in Fire and Polymers: Hazards Identification and Prevention. ACS Symposium Series 425, G.L. Nelson, ed. Washington, D.C.: American Chemical Society.

Kaplan, H.L., A.F. Grand, and G.E. Hartzell. 1983. Combustion Toxicology Principles and Test Methods. Lancaster, Pa.: Technomic Publishing Company, Inc.

Levin, B.C. 1992a. The development of a new small-scale smoke toxicity test method and its comparison with real-scale fire test. Pp. 257–264in Toxicology from Discovery and Experimentation to the Human Perspective, P.L. Chambers, C.M. Chambers, H.M. Bolt, and P. Preziosi, eds. New York: Elsevier Press.

Levin, B.C. 1992b. The development of a new small-scale smoke toxicity test method and its comparison with real-scale fire tests. Toxicology Letters 64/65:257–264.

Levin, B.C. 1994. A new approach for predicting the toxic potency of complex combustion mixtures. Proceedings of the American Chemical Society Division of Polymeric Materials: Science and Engineering 71:173–174.

Levin, B.C., A.J. Fowell, M.M. Birky, M. Paabo, A. Stolte, and D. Malek. 1982. Further Development of a Test Method for the Assessment of the Acute Inhalation Toxicity of Combustion Products. NBSIR 82-2532. Gaithersburg, Md.: National Institute of Standards and Technology.

Levin, B.C., M. Paabo, J.L. Gurman, and S.E. Harris. 1987a. Effects of exposure to single or multiple combinations of the predominant toxic gases and low oxygen atmospheres produced in fires. Fundamental and Applied Toxicology 9:236–250.

Levin, B.C., M. Paabo, J.L. Gurman, S.E. Harris, and E. Braun. 1987b. Toxicology interactions between carbon monoxide and carbon dioxide. Toxicology 47:135–164.

Levin, B.C., M. Paabo, and S.B. Schiller. 1991. A standard reference material for calibration of the cup furnace toxicity method for assessing the acute inhalation toxicity of combustion products. Journal of Research of the National Institute of Standards and Technology 96:741–755.

Levin, B.C., E. Braun, M. Navarro, and M. Paabo. 1995. Further development of the n-gas model: an approach for predicting the toxic potency of complex combustion mixtures. In Fire and Polymers, G. Nelson, ed. ACS Symposium Series. Washington, D.C.: American Chemical Society.

Norris, J.C. 1988. National Institute of Building Sciences Combustion Toxicity Hazard Test. Pp. 146–155 in Proceedings of the Joint Meeting of the Fire Retardant Chemicals Association and the Society of Plastics Engineers.

Purser, D. 1995. Smoke toxicity. Pp. 175–195 in Improved Fire- and Smoke-Resistant Materials for Commercial Aircraft Interiors: A Proceedings. National Materials Advisory Board report NMAB-477-2. Washington, D.C.: National Academy Press.

Roux, H.J. 1988. The NIBS SNITIX WG Program (for toxicity testing). Pp. 543–545 in Proceedings of the 37th International Wire and Cable Symposium, Reno, Nevada, November 15–17. Fort Monmouth, N.J.: U.S. Army Communications Electronic Command.

Appendix E

Biographical Sketches of Committee Members

ELI M. PEARCE (Chair) received a Ph.D. in polymer chemistry from the Polytechnic Institute of Brooklyn. He held various positions in industry—with DuPont's Carothers Research Laboratory, J.T. Baker Chemical Co., and Allied Chemical Corp. After serving as the director of the Dreyfus Laboratory of the Research Triangle Institute, he joined Polytechnic University as professor of chemistry and chemical engineering. Subsequently, he held positions as head of the Chemistry Department and dean of Arts and Sciences at Polytechnic. He is currently university professor and director of the Herman F. Mark Polymer Research Institute. His research interests center on polymer synthetic chemistry, structure/property relationships, degradation, flammability, and polymer blends.

BRUCE T. DEBONA received a B.S. from Rensselaer Polytechnic Institute and a Ph.D. in organic chemistry from the University of Pennsylvania. He is currently senior research associate in polymer chemistry in Allied Signal's Research and Technology Laboratory. His research interests include polymer chemistry, structure/property relationships, and the development of engineering thermoplastics, synthetic fibers, and polymeric composites.

FREDERICK L. DRYER received a B.S. in aerospace engineering from Rensselaer Polytechnic Institute and a Ph.D. in mechanical and aerospace engineering from Princeton University. He is currently professor of mechanical and aeronautical engineering at Princeton University. His research interests include fundamental combustion science, high-temperature chemistry of hydrocarbons, and fire-safety-related phenomena.

HOWARD W. EMMONS received an M.S. from Stevens Institute of Technology and an Sc.D. in engineering from Havard University. He is currently Gordon McKay Professor of mechanical engineering (emeritus) at Harvard University. His research interests include aerodynamics, combustion, gas dynamics, and fire science. He has been an innovator in the developing area of numerical characterization of fires and fire test methods. Dr. Emmons is a member of the National Academy of Sciences and the National Academy of Engineering.

TAKASHI KASHIWAGI received a B.S. and M.S. from Keio University (Japan) and a Ph.D. in aerospace and mechanical sciences from Princeton University. He joined the Fire Program at the National Institute of Standards and Technology (NIST) (then the National Bureau of Standards) in 1971. Dr. Kashiwagi is currently group leader of the materials fire research group of the Building and Fire Research Laboratory at NIST. His research interests include combustion modeling, polymer flammability, char formation, and flame spread in microgravity.

BARBARA C. LEVIN received a B.A. from Brown University and a Ph.D. from Georgetown University. She held post-doctoral and staff fellowships in the Laboratory of Molecular Biology at the National Institutes of Health before moving to the National Institute of Standards and Technology (NIST) where she conducted research on the toxicity of combustion products for 15 years. She is currently in the Biotechnology Division at NIST where she is examining the mutagenic effects of toxicants. Dr. Levin has conducted research concerning the toxicity of complex mixtures, toxicant suppressants, development of small-scale toxicity test methods, and validation of small-scale laboratory test results.

JAMES E. McGRATH received a Ph.D. in polymer science from the University of Akron. After several years in industry with ITT Rayonier, Goodyear Tire and Rubber, and Union Carbide, he joined the faculty at Virginia Polytechnic Institute and State University where he is the Ethyl Professor of polymer chemistry and director of the High Performance Polymeric Adhesives and Composites Center. He is a member of the National Academy of Engineering and the National Materials Advisory Board of the National Research Council. His research is in polymer synthetic chemistry, block and graft copolymers, high-temperature stable polymers and composites, and phosphorus-containing polymers.

JAMES M. PETERSON is a graduate of Wake Forest University and the California Institute of Technology. He has worked at The Boeing Company for over 25 years, and is currently a Boeing Technical Fellow. His primary responsibilities are fire safety of cabin interiors; FAA regulatory requirements, with special emphasis on fire-resistance standards for cabin furnishings; and general nonmetallic materials applications.

PATRICIA A. TATEM received a Ph.D. in physical chemistry from George Washington University. She is currently head of the Combustion Modeling and Scaling Section of the Navy Technology Center for Safety and Survivability at the Naval Research Laboratory. Dr. Tatem is involved in research and development work relevant to combustion with an emphasis on fire safety, including the modeling of the physical and chemical dynamics of the combustion process, scaling parameters, and intermediate and real-scale combustion experiments.